FUNCTIONAL NEUROLOGY

MOVEMENT 2019

BRAIN, BODY AND COGNITION

FUNCTIONAL NEUROLOGY

SERIES EDITORS -

GERRY LEISMAN, MD, PHD

F. R. Carrick Institute for Clinical Ergonomics,
Rehabilitation, and Applied Neurosciences (CERAN),
Garden City, New York, USA

and

JOAV MERRICK, MD

Medical Director,
National Institute of Child Health and Human Development,
Jerusalem, Israel

Movement 2019: Brain, Body and Cognition
Gerry Leisman and Joav Merrick (Editors)
2020. ISBN: 978-1-53616-714-6 (Hardcover)
2020. ISBN: 978-1-53616-715-3 (e-book)

Movement 2018: Brain, Body and Cognition
Joav Merrick and Gerry Leisman (Editors)
2019. ISBN: 978-1-53614-833-68 (Hardcover)
2019. ISBN: 978-1-53614-834-3 (e-book)

Neuroimmunity and the Brain-Gut Connection
Aristo Vojdani
2016. ISBN: 978-1-63483-969-3 (Hardcover)
2016. ISBN: 978-1-63483-970-9 (e-book)

Considering Consciousness Clinically
Gerry Leisman and Joav Merrick (Editors)
2016. ISBN: 978-1-63484-260-0 (Hardcover)
2016. ISBN: 978-1-63484-261-7 (e-book)

Neuroplasticity in Learning and Rehabilitation
Gerry Leisman and Joav Merrick (Editors)
2016. ISBN: 978-1-63484-305-8 (Hardcover)
2016. ISBN: ISBN: 978-1-63484-306-5 (e-book)

FUNCTIONAL NEUROLOGY

MOVEMENT 2019

BRAIN, BODY AND COGNITION

JOAV MERRICK
AND
GERRY LEISMAN
EDITORS

nova
Medicine & Health
New York

Library of Congress Cataloging-in-Publication Data

Names: Merrick, Joav, MD, MMedSci, DMSc, Medical Director, Health Services, Division for Intellectual and Developmental Disabilities, Ministry of Social Affairs and Social Services, Jerusalem, Israel; Division of Adolescent Medicine, KY Children's Hospital, Department of Pediatrics, Lexington, Kentucky, USA; National Institute of Child Health and Human Development, Jerusalem, Israel; Division of Pediatrics, Hadassah Hebrew University Medical Centers, Mt Scopus Campus, Jerusalem, Israel; School of Public Health, Georgia State University, Atlanta, Georgia, USA, editor. | Gerry Leisman, MD, PhD, Faculty of Health Sciences, University of Haifa, Mt Carmel, Haifa, Israel, editor.
Title: Movement 2019: Brain, Body and Cognition
Description: New York: Nova Science Publishers, [2019] | Series: Functional Neurology | Includes bibliographical references and index.
Identifiers: LCCN 2019954060 (print) | ISBN 9781536167146 (hardcover) |
ISBN 9781536167153 (adobe pdf)

Published by Nova Science Publishers, Inc. † New York

CONTENTS

INTRODUCTION

In: Movement 2019: Brain, Body and Cognition
Editors: Joav Merrick and Gerry Leisman
ISBN: 978-1-53616-714-6

Chapter 1

MOVEMENT, BRAIN, BODY AND COGNITION: A REFLECTION OF FUNCTIONAL CONNECTIVITY IN THE NERVOUS SYSTEM

Gerry Leisman, MD, PhD[1-3,*]
and Joav Merrick, MD, MMedSci, DMSc[4-7]
[1]Faculty of Social Welfare and Health Science, University of Haifa, Mt Carmel, Haifa, Israel
[2]National Institute for Brain and Rehabilitation Sciences, Nazareth, Israel,
[3]Department of Clinical Neurophysiology, Institute for Neurology and Neurosurgery, Universidad de Ciencias Médicas Facultad 'Manuel Fajardo' Ciudad de la Habana, Havana, Cuba,
[4]National Institute of Child Health and Human Development, Jerusalem, Israel,
[5]Department of Pediatrics, Mt Scopus Campus, Hadassah Hebrew University Medical Center, Jerusalem, Israel
[6]Kentucky Children's Hospital, University of Kentucky, Lexington, Kentucky, United States of America
[7]Center for Healthy Development, School of Public Health, Georgia State University, Atlanta, Georgia, United States of America

This book is based on the conference on "Movement and Cognition" held in July 2019 at Tel Aviv University in Israel, where an opportunity was provided for researchers, clinicians and practitioners from various disciplines to share their knowledge and experience in an academic environment. In this book you will find all the abstracts from this conference gathered in one publication. We believe that movement facilitates

[*] Corresponding Author's Email: g.leisman@edu.haifa.ac.il.

cognition throughout the life span and hope that this book will be of interest to both researcher, clinicians, practitioners and other people who are interested in the issue of movement and the brain.

Introduction

The brain and its relationship to movement is apparent when one considers the classic model of the reflex arc in which a sensory stimulus interacting through the spinal-cord will result in a response. What is less apparent are the evolutionary origins of cognitive motor interaction. Why did we as humans develop such systems and how are these systems related to the activities of our daily lives and how do such systems develop in childhood and decrement as a consequence of the aging process? We are very much aware of how neonatal influences develop motor and cognitive competencies, and we can witness how toddlers can suddenly become motorically adept. We accept that professional gymnasts are over with their careers as performers by 25 years of age if not before and that elite sports athletes move to the world of coaching or another career by the time they reach 40 years of age. We are all aware of the motor deficits associated with the elderly and on changes as we advance in years.

The third International Conference on "Movement and Cognition" met 22-24 July 2019 on the campus of the University of Tel Aviv in Israel. Previously the conference had taken place at Harvard Medical School in 2018 and prior to that at the University of Oxford in 2017.

The congress brought together individuals; clinicians, researchers, and engineers as well as those from the world of dance, and somatic body work under one roof with university sponsorship. It was in this environment in which a furtive interchange occurred between individuals speaking different languages yet working on similar applications from developmental issues, to rehabilitation, training of elite sports athletes to applications in human aging.

The gathering has allowed us to begin to build a model of the human being in which the brain is the control center, directing the movements of the body but perhaps more comprehensively than we normally envision the process. Our brain signals heartbeat and digestion. When muscles are instructed to contract skeletal function is controlled. With movement disorders, such as gate freeze there exists incapacity of movement initiation or control. Most interventions center on the restoration of brain function. For example, most medications for Parkinson's disease target cerebral neurotransmitters. Such tactics apparently make sense. By repairing the control center, that center can return to its supervisory role. But new developments in neuroplasticity and in the treatment of movement impairments show that such a top-down, brain-focused approach might not be the most optimized pathway for ameliorating such problems. We have learned that we may need to rethink our brain-first model (1, 2).

The continuing debate about the definition of death in part results from the technological deconstruction of dying. We in the past had it simple. A construction of a binary state of living or dead allowed an easy determination, but today we are more likely dealing with something more akin to a "syndrome," consisting of a cluster of related attributes, with a probabilistic diagnosis (2). We are now required to determine how many of these features are necessary before we begin to "proclaim" someone as "dead," just as psychiatrists must decide how many psychiatric traits are required before making a diagnosis of "schizophrenia." Electroencephalograms can only resolve a termination of electrical activity on the brain surface, and not in the deeper structures. It can also not establish if the electrically inactive brain tissue can regenerate activity. Many of those who are diagnosed as brain dead in fact have clear evidence of functioning midbrains and brainstems, and are not necessarily irreversible (3, 4). A key argument in favor of whole brain death criteria over neo-cortical death, is that the brain provides integrative functions that the body needs to survive. This has also been shown to be fallacious since patients meeting the current clinical criteria for whole brain death have survived for years.

The relevance of the forgoing to the purpose of the conference is to provide a framework for a better understanding of the integrators of brain and nervous system function. Treating brain disorders, educating children in the world's school systems, producing elite athletes, reducing ADD/ADHD and autism spectrum disorders, using dance as a treatment tool, seems to require more than brain-centric therapies. Interaction with our environments by movement promotes brain plasticity, and movement is key to optimizing brain function. The success of constraint-induced therapy and high-intensity exercise in treating neurological disease and dysfunction or even developmental delays suggests that the control center model of the brain is incorrect, that the brain is not the underlying cause of these disorders. Rather, the brain and body function, learn and adapt together. So treatment should address both physical actions and cognitive processes. And the treatment stimulus must be demanding enough to require adaptations from the brain and body that will allow them to regrow, heal and improve together.

What the abstracts are largely presenting in this volume that represents the interdisciplinary conference relates to brain and nervous system efficiency that we have described more fully elsewhere (5-9). Higher abilities in cognitive tasks are associated with more efficient neural processing and not necessarily a particular brain region that is involved in that processing. Whereas intuitively, we would expect higher performance to correlate with more activity, for the cerebral cortex the contrary is the case. Higher performance in several tasks, including verbal (10), numeric, figural, and spatial reasoning (11, 12) is consistent with the reduced consumption of energy in several cortical areas. This phenomenon has also been studied with EEG techniques in different frequency bands. The amount of a background power (7.5–12.5 Hz) decreases during cognitive activity compared with a resting state. This decrease has been observed to correlate with higher performance in subjects with higher IQ scores (7) or with higher performance after training, indicating a more efficient processing strategy for a cognitive task (13). Most of these studies come from the psychological literature, focusing mainly on the domain of intelligence.

The point is that greater activity is demonstrated with less proficiency and vice versa. The function of childhood neurological development is precisely to facilitate the creation of localized function and it is dynamic. It can be changed and is therefore plastic. This localization of function is not the explanation of a process, but rather the end-result of training. The efficiency of cognitive function is directly a consequence of the effectiveness of networks that now can be measured. Fewer brain regions necessary to accomplish a single task in one individual compared to another for the same task is a measure of efficiency.

These networks, active during learning and problem solving of all kinds, are plastic and can be changed as a direct consequence of experience and training.

In attempting to apply graph theory concepts to child and adolescent neurocognitive performance to create a fundamental change in the educational training and evaluation paradigm, we can characterize the organization & development of large-scale brain networks using graph-theoretical metrics as represented in figure 1 below.

What we can learn from the characterization, organization and development of large-scale brain networks in children, using graph-theoretical metrics, is that small-world networks are characterized by an increased clustering coefficient, or an average node-to-node distance (also known as average shortest path length) and a decreased characteristic path length (and represented in figure 1). Functional brain networks in children and young-adults show small-world properties. In mathematics, physics and sociology, a small-world network is a type of mathematical graph in which most nodes are not neighbors of one another, but most nodes can be reached from every other node by a small number of steps. Specifically, a small-world network is defined to be a network where the typical distance L between two randomly chosen nodes (the number of steps required) grows proportionally to the logarithm of the number of nodes N in the network that is (14):

$$L \alpha \log N$$

In the context of a social network, this results in the small world phenomenon of strangers being linked by a mutual acquaintance. Many empirical graphs are well modeled by small-world networks. Social networks, the connectivity of the Internet, Wikipedia, and gene networks all exhibit small-world network characteristics (15).

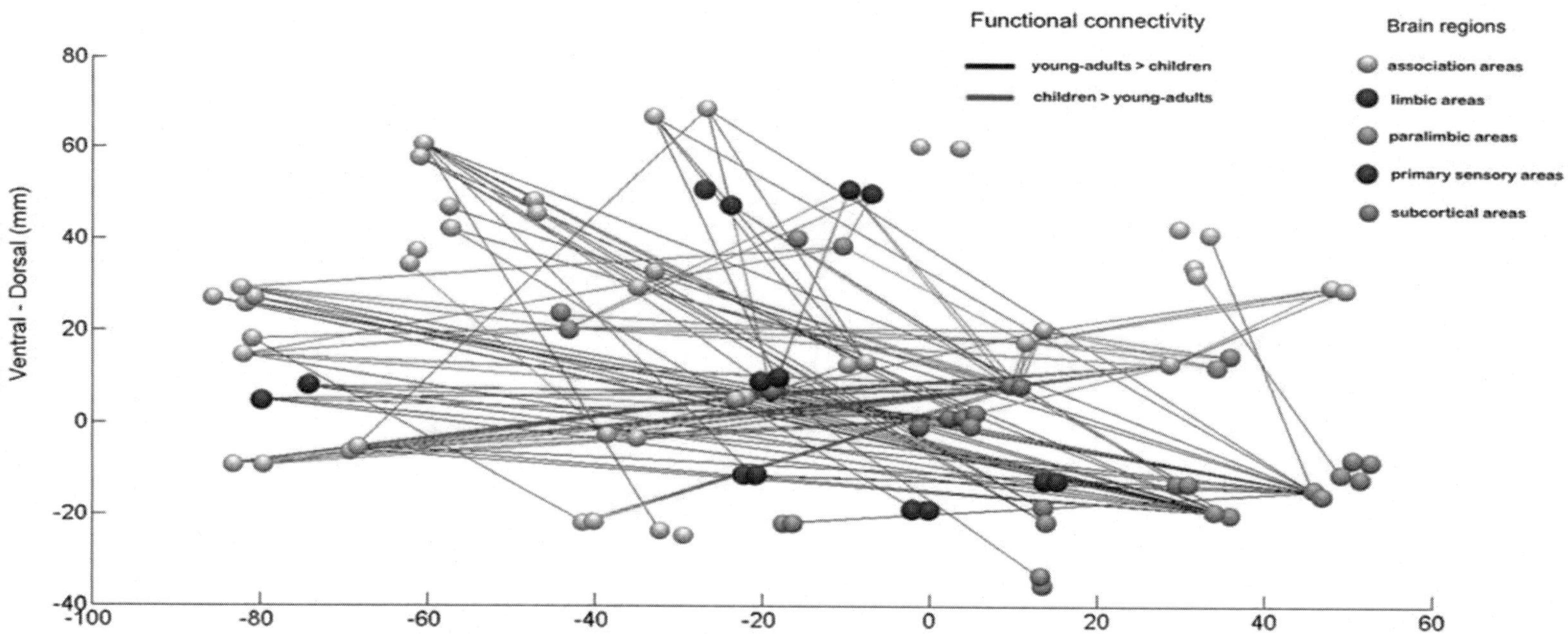

Figure 1. Functional connectivity along the posterior-anterior and ventral-dorsal axes showing increased subcortical connectivity (●), decreased paralimbic connectivity (●) in children, compared to young-adults.

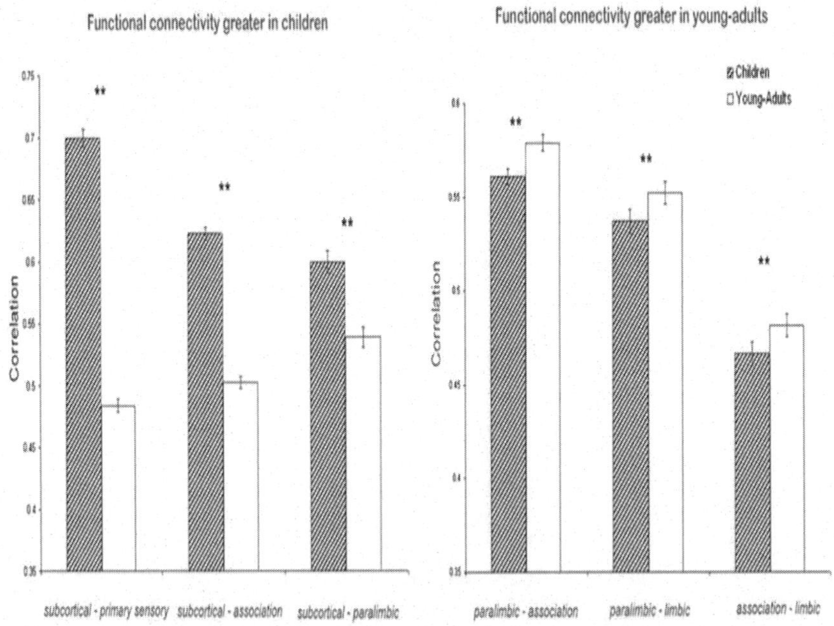

(A)

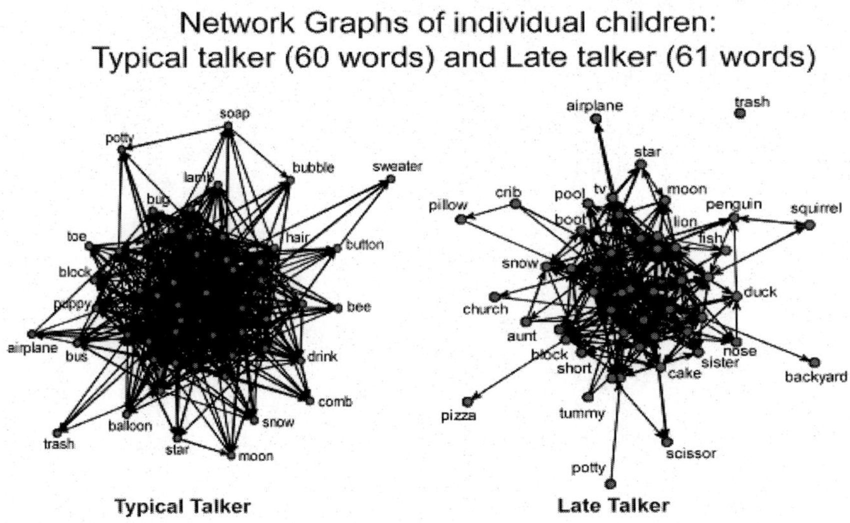

	Child 1 (TD) 60 words (40%, 17mo)	Child 2 (LT) 61 words (10%, 24mo)	Random acquisition network
clustering coefficient	0.641	0.392	0.485
geodesic dist.(mean)	1.526	2.863	2.049
In-degree (median)	25	9	13.95

(B)

Figure 2. Characterization, organization & development of large-scale brain networks in children using graph-theoretical metrics. (A) The graph on the left is a typically developing (TD) child (17 mo., 40%) and the graph on the right is of an at-risk, late-talker (LT) (24 mo., 10%). (B) The network of the TD child includes the 60 words in the child's productive vocabulary and the network of the at-risk LT child includes the 61 words in the child's productive vocabulary. The apparent visual differences in the networks are supported by the differences in the corresponding table, with the typical talker's network showing higher clustering coefficient and higher median in-degree, but lower geodesic distance, than the LT.

These findings suggest sub-networks of densely connected nodes, connected by a short-path. Functional connectivity networks of brain from EEG (16, 17) as well and MEG (17) have also been shown to possess small-world architecture. Large-scale brain networks in 7-9-year-old children show similar small-world, functional organization. Functional brain networks in children show lower levels of hierarchical organization compared to young-adults. Children and young-adults possess different interregional connectivity patterns, stronger subcortical-cortical connectivity's in young adults and weaker cortico-cortical connectivity's in children. Large-scale brain connectivity involves functional segregation and integration, stronger short-range connections in children, and stronger long-range connections in young-adults. In taking this concept further, we note that represented in figures 2(A) and (B) below are representations of functional connectivity along the posterior-anterior and ventral-dorsal axes showing elevated subcortical connectivity and decreased para-limbic connectivity in children, compared to young-adults. We can clearly demonstrate that the wiring and connectivity's of young children are significantly different than teenagers and beyond and the change in organization of these connectivity's directly speaks to the issue of optimization of pathways and is a direct consequence of training and therefore of education.

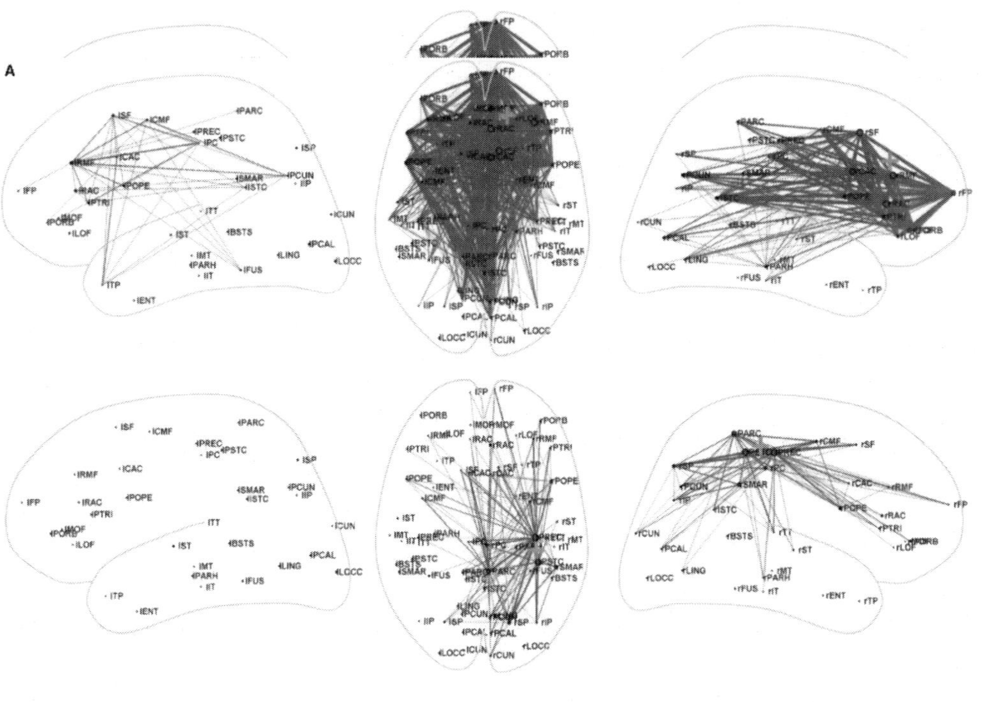

(B)

Figure 3. Demonstration of computational modulations in connectivity resulting from lesions in the (A) frontal cortex including motor control and executive function areas and (B) sensorimotor cortex. Red lines indicate strength in connectivity. Note the widespread disruption caused by lesion in prefrontal cortex compared with relatively constrained, intra-hemispheric changes resulting from a lesion of the sensorimotor cortex.

The issue of networks sub-serving neuroplasticity in general and cognitive motor-interaction in particular can be demonstrated in figure 3 below where damage or dysfunction in frontal and prefrontal neocortical regions has a significantly greater overall deleterious effect on brain networks than does a lesion in the somatosensory cortex. This again supports the notion that a fundamental understanding of brain network architecture I are important than site of a lesion and that networks can be trained, retrained, or rerouted.

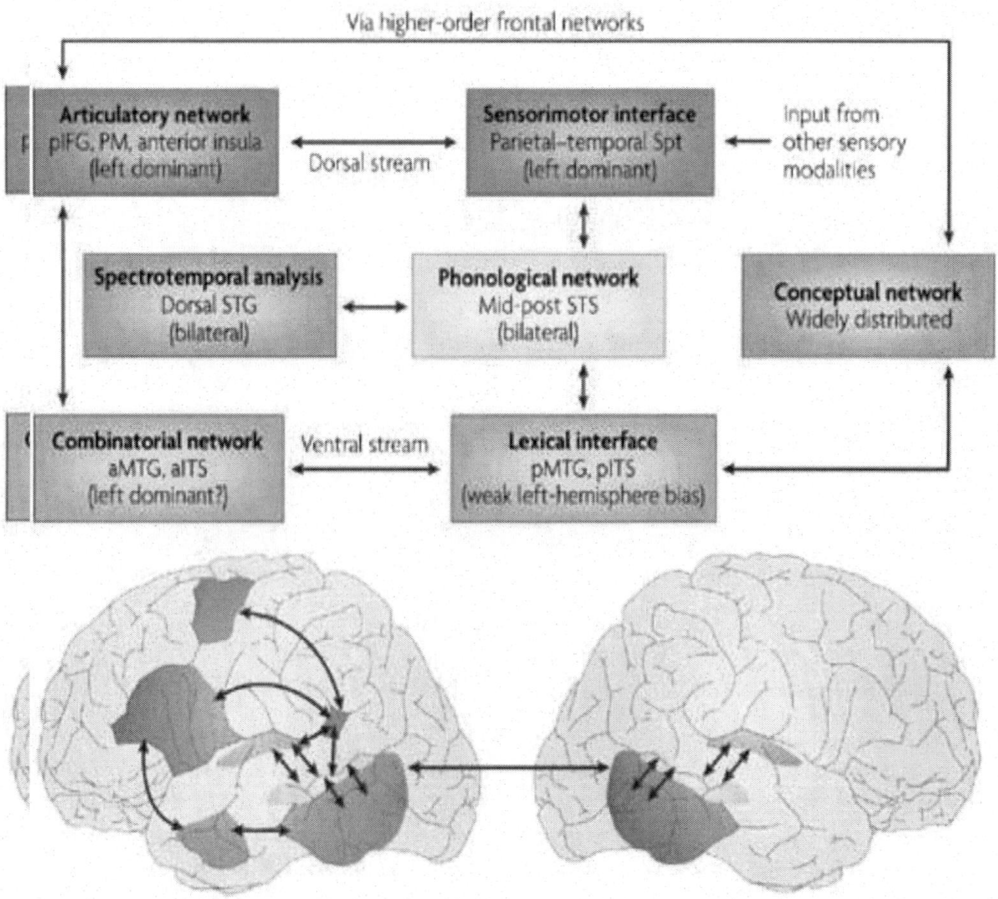

Figures 4. (A) Bye to the good old days: No more receptive functions in one (Wernicke's) area. Multiple stream models are more likely. Receptive language functions are organized into multiple self-organizing simultaneously active networks. (B) Grounded meaning indicates that the meaning of words and sentences have been claimed to be "embodied".

It has been thought since the time of both Broca and Wernicke that there exists a high degree of localization of function with an area anterior to the Sylvian fissure of the temporal lobes being responsible for expressive language and Wernicke's area responsible for comprehension. Today we better understand that there no longer exists the localization of receptive functions in a given area of the brain. Multiple stream models are more likely. Receptive language functions are organized into multiple self-organizing simultaneously active networks. It appears also as represented in Figs. 4 that the meaning of words and sentences has been grounded indicating that there is an "embodiment" of meaning in brain networks as previously described.

Given the nature of brain networks and the need to provide a comprehensive and comprehensible language to understand how they are constructed, the conference has been attempting to provide that language and is represented in the abstracts that appear in this edition.

Conclusion

We conclude that that the brain evolved for the control of action rather than for the development of cognition *per se* the latter being a consequence of agency (2). Currently employed descriptions of brain–behavior relationships actually limit how we those relationships are both understood and investigated, thereby limiting the both the questions we ask and the value of the "answers". Many concepts are so broad and too far-reaching rendering them scientifically insignificant. "Executive function" is one such commonly used term. As the construct is increasingly focal in neuroscience research, defining it clearly is critical. A more operational term would allow us to understand the relationship that we human have with our environment better as well as provide a language to share ideas between disciplines (2, 18). Such a change in operationalized terminology such as continuous sensorimotor interaction with the environment would yield better results. The chief function of "executive function" is optimized environmental interaction, which would be most effectively seen in our ability to anticipate and control behavior with both implicit and explicit mechanisms. This, in turn, would allow us to understand how to conduct an orchestra or play in a string quartet, dance w tango with partner, play football, "re"-habilitate as well as habilitate or educate. Procedural memory exemplified by motor skill learning contributes to knowledge, thereby grounding that knowledge in sensorimotor anticipation, thereby directly linking movement to thought and "embodied cognition" (2,18). This text and the abstracts of the conference that reflects the importance of Movement, Brain and Cognition are to be found in these pages.

References

[1] Israely S, Leisman G, Carmeli E. Neuromuscular synergies in motor control in normal and post stroke individuals. Rev Neurosciences 2018;29(6):593-612.

[2] Melillo R, Leisman G. Neurobiological disorders of childhood: an evolutionary approach. New York: Springer, 2009.

[3] Machado C, Leisman G. Towards an effective definition of death and disorders of consciousness. Rev Neurosciences 2009;20(3-4):147-50.

[4] Leisman G, Koch P. Networks of conscious experience: computational neuroscience in understanding life, death, and consciousness. Rev Neurosciences 2009;20(3-4):151-76.

[5] Ertl JP, Schafer EWP. Brain response correlates of psychometric intelligence. Nature 1969:223(5204);421-2.

[6] Grabner RH, Stern E, Neubauer AC. When intelligence loses its impact: neural efficiency during reasoning in a familiar area, Intern J Psychophysiology 2003;49(2):89–98.

[7] Grabner RH, Fink A, Stipacek A, Neuper C, Neubauer AC. Intelligence and working memory systems: evidence of neural efficiency in alpha band ERD. Cog Brain Res 2004:20(2);212–25.

[8] Gilchriest J. A method for quantifying visual search scanpath efficiency. Functional Neurology Rehabilitation Ergon 2011;1(2):181-96.

[9] Leisman G, Machado C, Melillo R, Mualem R. Intentionality and "free-will" from a neurodevelopmental perspective. Front Integrat Neurosci 2012;6:36. doi: 10.3389/fnint.2012.00036.

[10] Parks RW, Loewenstein DA, Dodrill KL, Barker WW, Joshii F, Chang JY, et al. Cerebral metabolic effects of a verbal fluency test: a PET scan study. J Clin Exp Neuropsychol 1988;10:565–75.

[11] Lamm C, Bauer H, Vitouch O, Gstattner R. Differences in the ability to process a visuo-spatial task are reflected in event-related slow cortical potentials of human subjects, Neurosci Lett 1999;269(3):137–40.

[12] Vitouch O, Bauer H, Gittler G, Leodolter M, Leodolter U. Cortical activity of good and poor spatial test performers during spatial and verbal processing studied with slow potential topography, Intern J Psychophysiol 1997;27(3):183–99.

[13] Neubauer, AC, Grabner, RH, Freudenthaler H, Beckmann H, Jens F, Guthke J. Intelligence and individual differences in becoming neurally efficient. Acta Psycholgica (Amsterdam) 2004;116(1):55– 74.

[14] Watts, Duncan J, Strogatz SH. Collective dynamics of 'small-world' networks. Nature 1998;393:440-2.

[15] Howard N, Leisman G. DIME (Diplomatic, information, military and economic power) effects modeling system: Applications for the modeling of the brain. Funct Neurol Rehabil Ergon 2013:3:2-3.

[16] Leisman G. Brain networks, plasticity, and functional connectivities inform current directions in functional neurology and rehabilitation. Funct Neurol Rehabil Ergon 2011;1(2):315-56.

[17] Stam CJ. Functional connectivity patterns of human magnetoencephalographic recordings: a 'small-world' network? Neurosci Lett 2004;355(1-2):25-8.

[18] Leisman G, Moustafa AA, Shafir T. Thinking, walking, talking: integratory motor and cognitive brain function. Front Public Health 2016;4:94.

SECTION ONE: ABSTRACTS

Abstracts from the 2019 International Conference on Movement and Cognition at the Tel Aviv University, Israel, July 22-24, 2019.

In: Movement 2019: Brain, Body and Cognition
Editors: Joav Merrick and Gerry Leisman
ISBN: 978-1-53616-714-6

Chapter 2

DEFORMITY OF FEET, KNEES, HIPS, PELVIS AND SPINE CAUSED BY "MINIMAL BRAIN DYSFUNCTIONS"

Tomasz Karski*, Jacek Karski and Marian Domagala
Vincent Pol University in Lublin, Medical University in Lublin and Medical Center, Laszczów, Poland

Keywords: minimal brain dysfunction, deformities

Introduction: In Poland almost 7% of children are born with minimal brain dysfunction (MBD). The causes of MBD are in pre-natal, perinatal and post-natal period of child's life. Consequently, secondary changes of the locomotors system can be observed. In children with MBD we observe the sub-spasticity of muscles and very frequently laxity of joints because of changed properties of collagen,

Material: This article provides observations from the years 1995-2018. The material consists of 998 children and youths aged from 2-18 years. The result of MBD can be: valgus deformity of the feet, hyperextension/recurvation of the knees, anterior tilt of the pelvis, and hyperlordosis of the lumbar spine. Valgus or plane valgus deformity of the feet with the following deformity: 1) while walking we need in every step the dorsal flexion of the feet, 2) in the cases of MBD it is shortening of the Achilles tendon and m. triceps surae and in case that the stability of joints is proper/normal – the child's walking is on equines position of toes (toes walking) 3) if the shortening of Achilles tendon and m. triceps surae coexist with laxity of joints, the required dorsal flexion is made in prone position of the feet, by every step, 4) such repeated "dorsal flexion in prone position" after some years give "full fixed valgus, or "plane - valgus deformity" of the feet.

* Corresponding Author's Email: t.karski@neostrada.pl.

Recurvation and valgus deformities of the knees. Recurvation of the knees is very often accompanied by valgus deformity of the feet. Such deformity of the knees is the effect of a shortening of the Achilles tendon and m. triceps surae. In lecture authors explain these connections. Also to valgus deformity of knees lead the improper sitting, which will be presented. Anterior tilt of the pelvis and hyperlordosis of the lumbar spine. In children with MBD, exist spasticity or sub spasticity of m. rectus, one part of the m. quadriceps. It is the cause an anterior tilt of the pelvis and hyperlordosis of lumbar spine. Not treated this deformity lead at adults people to “low back pain syndrome”.

Conclusion: In Poland from 7-11% of children and youth have the symptoms of minimal brain dysfunctions (MBD). In result of MBD there is shortening of tendons and muscles and if contemporary exist laxity of joints - and can develop feet valgus or plano–valgus deformity, recurvation of knees, anterior tilt of pelvis with hyperlordosis of lumbar spine. Early treatment of all above mentioned deformities in childhood is the best prophylactics of pain syndromes in feet, knees, hips and spine in adults. In the treatment, the best method is stretching exercises. Only 5% of patients need surgery.

In: Movement 2019: Brain, Body and Cognition
Editors: Joav Merrick and Gerry Leisman
ISBN: 978-1-53616-714-6

Chapter 3

BIOMECHANICAL ETIOLOGY OF THE SO-CALLED IDIOPATHIC SCOLIOSIS (1995-2007): NEW CLASSIFICATION, NEW TREATMENT AND RULES OF CAUSAL PROPHYLACTICS

Tomasz Karski*
Vincent Pol University, Lublin, Poland

Keywords: scoliosis, adolescent idiopathic scoliosis

Introduction: The biomechanical etiology of the so-called idiopathic scoliosis [adolescent idiopathic scoliosis (AIS)] will be discussed on the basis of research conducted.

Material: 2,500 cases aged 2-60 years old with control group of 505 persons over the 1995-2007 period. Research with these cases has resulted in a new classification – three groups and four types.

1) "S" I etiopathological (epg) scoliosis Double curves. Gibbous of the right side. Influenced by the "gait" and the permanent "standing at ease on the right leg". Stiff spine. 3D. Progression.
2A) "C" II/A epg scoliosis. Influenced by the permanent "standing at ease on the right leg". One curve. Flexible spine. 1D. No or slight progression.

* Corresponding Author's Email: tmkarski@gmail.com.

2B) "S" II/B epg scoliosis. Influenced by the permanent "standing at ease on the right leg", plus - laxity of joints or/and incorrect exercises in previous treatment. Flexible spine. 2D or mix. Moderate progression.

3) "I" III epg scoliosis. Influenced by the "gait" only. Stiff spine. No curves or small. No progression. No included till now (2007) to scoliosis group.

Physiotherapy. All previous extensions, "muscles strengthening exercises" were incorrect. They caused only bigger curves, bigger rib hump and stiffness of spine. All stretching exercises for spine and hips are proper for treatment and for prophylaxis. Very important in therapy are: karate, taekwondo, aikido and standing 'at ease' on the left leg.

Conclusion: The etiology of the so-called idiopathic scoliosis is strict biomechanical. There are three groups and four types of scoliosis – connection with "standing" and with "gait". Every type of scoliosis starts at 2-3 years of life. In therapy and in causal prophylaxis are important - the stretching exercises introduced very early (3-5 years of age) and standing on the left leg.

In: Movement 2019: Brain, Body and Cognition
Editors: Joav Merrick and Gerry Leisman
ISBN: 978-1-53616-714-6

Chapter 4

PELVIC SCHEMA AND GRAPHIC-METRIC REPRESENTATION IMPROVE FOLLOWING DYNAMIC NEURO-COGNITIVE IMAGERY TRAINING IN PEOPLE WITH PARKINSON'S DISEASE

***Amit Abraham*[1,2,*]*, Ariel Hart*[1]*, Ruth Dickstein*[3]**
***and Madeleine E Hackney*[1,4]**
[1]Department of Medicine, Division of General Medicine and Geriatrics, Emory University School of Medicine, Atlanta, Georgia, United States of America
[2]Department of Kinesiology, University of Georgia, Athens, Georgia, United States of America,
[3]Department of Physical Therapy, Faculty of Social Welfare and Health Sciences, University of Haifa, Haifa, Israel
[4]Atlanta Department of Veterans Affairs, Center for Visual and Neurocognitive Rehabilitation, Atlanta, Georgia, United States of America

Keywords: Parkinson's disease, pelvis, body schema, mental imagery, dynamic neuro-cognitive imagery

Objective: Body schema (i.e., the mental representations of the body), vital for motor and cognitive functions, is often distorted in people with Parkinson's disease (PD). Deficits in body, and especially pelvic, schema can further exacerbate motor and cognitive deficits associated with PD. Such deficits, including those in graphic and metric misjudgments, can manifest in drawing tasks. Mental imagery is a recommended

[*] Corresponding Author's Email: amit.abraham@emory.edu.

approach for PD rehabilitation with potential for ameliorating body schema. The goal of the current study was to investigate the effect of a two-week dynamic neuro-cognitive imagery (DNI) training versus in-home learning and exercise control (learning/exercise) on pelvic schema and graphic representation (i.e., drawing height and width).

Method: Twenty participants with idiopathic PD (Hoehn & Yahr I-III; M age: 65.75+10.13) were randomly allocated into either a DNI or a learning/exercise group. Participants were asked to complete the "Draw your pelvis" test in which they drew their pelvis at pre- and post-intervention. Drawings were assessed for pelvic schema score and drawing dimensions (i.e., height and weight) by trained raters. The DNI intervention focused on anatomical and metaphorical imagery of pelvic anatomy and biomechanics.

Result: No difference ($p > .05$) was detected at baseline between drawn pelvis height and width. Following intervention, improvements were greater in the DNI group for pelvic schema ($p < .01$), drawn pelvic width ($p < .05$) and width-height difference ($p < .05$).

Conclusion: This study suggests that DNI could serve as a rehabilitation path for improving body schema in people with PD. Future studies should explore DNI mechanisms of effect and the effect of enhanced pelvic schema on motor and non-motor deficits in this population.

In: Movement 2019: Brain, Body and Cognition
Editors: Joav Merrick and Gerry Leisman
ISBN: 978-1-53616-714-6

Chapter 5

THEN MECHANISM OF LEVODOPA IN PARKINSON'S DISEASE: A DIRECTED FUNCTIONAL CONNECTIVITY PATHWAY ANALYSIS

Gadi Goelman[1,*], Rotem Dana[1,2], Filip Ružickac[3,4], Ondrej Bezdicekc[3] and Robert Jech[3,4]
[1]Department of Neurology, Hadassah Hebrew University Medical Center, Jerusalem, Israel,
[2]Edmond and Lily Safra Center for Brain Sciences (ELSC), Hebrew University of Jerusalem, Jerusalem, Israel,
[3]Department of Neurology and Center of Clinical Neuroscience, First Faculty of Medicine and General University Hospital, Charles University, Prague, Czech Republic
[4]Na Homolce Hospital, Prague, Czech Republic

Keywords: MRI, functional connectivity, levodopa, Parkinson's disease, wavelet analysis

Objective: To understand the mechanism by which levodopa affects the motor system in Parkinson's disease (PD).

Method: A novel multivariate functional connectivity method that deduces directed functional pathways based only on uniform hemodynamic response across the cortex, was used with resting-state fMRI data of PD patients at OFF and ON medication states (N = 37) and of healthy controls (N = 41). Analysis was restricted to the motor system excluding the cerebellum.

[*] Corresponding Author's Email: gadig@hadassah.org.il.

Result: The majority of directed pathways were bottom-up pathways with top-down pathways only 5%, 4% and 9% for control, PD-OFF and PD-ON states. Most bottom-up pathways included the module: 'S1→M1→SMA' (S1/M1, the primary sensory/motor cortex; SMA, the supplementary motor area) with reduced numbers in PD OFF and ON states versus control (32% and 35%). The main relative effect of levodopa was in top-down pathways cooperating S1, M1, the anterior and dorsolateral prefrontal cortex. Great reduction in the number of these pathways versus controls was observed in PD-OFF state (25%) but only of 57% in PD-ON state. The method enabled detailed observations: For example, the pathways 'DLPFC→ aPFC→S1→M1' were observed equally in all groups but the pathways 'aPFC→DLPFC→S1→M1' were observed only in the controls and the PD ON state groups. The former is related to goal directed while the latter to automatic movements.

Conclusion: Our results support viewing PD as a network disorder with reduced bottom-up pathways in line with the classical models of PD. The effect of levodopa was to restore top-down pathways especially from the PFC.

In: Movement 2019: Brain, Body and Cognition
Editors: Joav Merrick and Gerry Leisman
ISBN: 978-1-53616-714-6

Chapter 6

ONLY THREE FINGERS WRITE, BUT THE WHOLE BRAIN WORKS: A HIGH-DENSITY EEG STUDY SHOWING ADVANTAGES OF DRAWING OVER TYPING FOR LEARNING

Ruud van der Weel **[*]** ***and Audrey van der Meer***
Department of Psychology and Developmental Neuroscience Laboratory,
Norwegian University of Science and Technology (NTNU), Trondheim, Norway

Keywords: cognitive neuroscience, electroencephalography (EEG), electrophysiology of handwriting, teaching

Objective: Are different parts of the brain active when we type on a keyboard as opposed to when we draw visual images on a tablet?

Method: Electroencephalogram (EEG) was used in young adults to study brain electrical activity as they were typing or describing in words visually presented Pictionary™ words using a keyboard, or as they were drawing pictures of the same words on a tablet using a stylus. Analyses of temporal spectral evolution (time-dependent amplitude changes) were performed on EEG data recorded with a 256-channel sensor array.

Result: We found that when drawing, brain areas in the parietal and occipital regions showed event related desynchronization activity in the theta/alpha range. Existing literature suggests that such oscillatory neuronal activity provides the brain with optimal conditions for learning. When describing the words using the keyboard, upper

[*] Corresponding Author's Email: ruud.weel@ntnu.no.

alpha/beta/gamma range activity in the central and frontal brain regions were observed, especially during the ideation phase. However, since this activity was highly synchronized, its relation to learning remains unclear.

Conclusion: We concluded that because of the benefits for sensory-motor integration and learning, traditional handwritten notes are preferably combined with visualizations (e.g., small drawings, shapes, arrows, symbols) to facilitate and optimize learning.

In: Movement 2019: Brain, Body and Cognition
Editors: Joav Merrick and Gerry Leisman
ISBN: 978-1-53616-714-6

Chapter 7

TOWARDS AN EMBODIED LANGUAGE FOR EVERYONE: A DANCE WORKSHOP

Owen Allen*
Physiotherapist, Artistic Director, Phoenix Functions, Atherton, Qld, Australia

Keywords: dance, embodied inclusive language

Objective: Participants experience an accessible 2dance (contemporary dance theatre) program that optimizes learning, physical skill training, social engagement and collaboration, creativity and play, performance and mastery in an inclusive context. Participants are introduced to a notion of a future in which there is universal sophisticated movement language.

Method: An active workshop based on the inclusive dance design method developed by the author. It is based on an ontological and phenomenological framework i.e., the experience and exploration of being a dancer, being authentic, being a movement student, being a collaborator, and being a performer. For the purposes of the work, a dancer is anyone who has entered the dance space. The method is derived from neuroplasticity optimization training characteristics including explicit, implicit and creative learning modalities. Utilizing the participant's own creative ability to design from the simplest movement expressions, and their collaboration with other participants, they are introduced to a practice method that is inherently empowering and adaptable to diverse performer abilities and performative purposes. The participant is introduced to designing a complete biomechanical and embodied language as an experimental method in movement training and performance.

* Corresponding Author's Email: owen@phoenixfunctions.com.au.

Result: Participants will have experienced movement learning through their own creative ability. They will have danced. They will have begun to explore the possibility of designing a universal and sophisticated language of movement including defining a basic unit of a biomechanical language, such as might be termed a moveme (moo-veem), as per the linguistic terms phoneme and grapheme.

Conclusion: Inclusivity in dance can be greatly enhanced by a working method that begins with what the dancer can bring to the floor and with empowering the dancers own creative and collaborative abilities. Such method starts with the dancer becoming clear about the simplest movements that could be thought of as beginning to understand something of what may become the basic units of an embodied language.

About the Author

Owen Allen is an Australian physiotherapist of 38 years' experience and 8 years of dance culminating in the design of an inclusive dance program, rEvolve.

Email: http://phoenixfunctions.com.au/rEvolve/index.html.

As a mature dancer, his 8 years of dance practice has been with thanks to the mentoring from North Queensland regional dancers, and professional dancers in residency to the Cairns Centre of Contemporary Arts. He has been involved in 2 DANscienCE Festivals and the inaugural BOLD Festival, in Australia, and these have been a significant contribution to his practice and the aspiration of the very real possibility of dance for everyone.

In: Movement 2019: Brain, Body and Cognition
Editors: Joav Merrick and Gerry Leisman
ISBN: 978-1-53616-714-6

Chapter 8

NOT JUST A BODY LANGUAGE: FACILITATED VS CHOREOGRAPHED METHODS SO THAT EVERYONE CAN DANCE AND EVERYONE CAN DANCE WITH ANYONE

Owen Allen*
Physiotherapist, Artistic Director, Phoenix Functions, Atherton, Queensland, Australia

Keywords: dance embodied inclusive language community

Presentation type: Demonstrative and inclusive (adults with disabilities) the workshop will be a version of the workshop that will be demonstrated during the Movement Is Conference. (20 or 60 minute session).

Technical requirements: Open floor space (dance flooring not required)

Presentation: Utilizing the participant's own creative ability to design from the simplest movement expressions and in collaboration with other participants, a work and performance design method that is inherently empowering and adaptable to diverse performer abilities and performative purposes, is presented. The work is based on experimenting from the simplest units of movement as embodied conversations toward the evolution of a sophisticated biomechanical language that works as a daily social expression through to the complexity of a great poem or novel. Owen proposes calling the basic unit of such a language a 'mov-eme' in line with linguistic standards. In wondering how the basic units of such a language, would be, he invites people (dancers) into an experiment that can evolve over generations towards the development of a language of body (not just a body language).

---

* Corresponding Author's Email: owen@phoenixfunctions.com.au.

In: Movement 2019: Brain, Body and Cognition
Editors: Joav Merrick and Gerry Leisman
ISBN: 978-1-53616-714-6

Chapter 9

ADVANCES IN PHARMACOTHERAPY OF TRAUMATIC BRAIN INJURY: NEURAL SUPPORT VERSUS NEURAL PROTECTION

Theresa Onazi*
University of Debrecen, Debrecen, Hungary

Keywords: magnesium sulfate, fetal neurologic injury, prevention

Objective: Magnesium sulfate, a widely known pharmaceutical is commonly administered in obstetric practice for protection against fetal neurologic injury. Although researches and meta-analysis have elicited its efficacy, it is also toxic to the neonate. In the same vein, preliminary research suggests epidural spinal cord stimulation can increase heart rate and blood pressure in patients with cervical spinal cord injury who suffer from orthostatic hypotension. This study will be focused on the effects of pharmacotherapeutic interventions in traumatic brain injury, with a focus on in neural support versus neural protection roles, and promising paths of the nervous system's adaptability.

Method: A thorough search of the United States National Library of Medicine resulted in a myriad of clinical trials, articles and expert opinions regarding the historical evolution of pharmacotherapeutic interventions in traumatic brain injury, as well as what is known about neuroprotection and neural support thus far. Pharmacotherapeutic interventions that resulted in positive and negative effects, post traumatic brain injury, were focused on.

* Corresponding Author's Email: enekaith@yahoo.com.

Result: There is incomplete comprehensive understanding of the ever elusive nervous system, especially in the setting of traumatic brain injury. Minimal data studies the effects of pharmacotherapy in traumatic brain injury, detailing its potential neurosupport and neural protection role.

Conclusion: Connections between the adaptability of the nervous system post insult and intervention are increasingly being recognized. Traumatic brain injury has a wide etiological background. Internal factors such as genetics, and external factors such as pharmacotherapy interestingly play huge roles separately, additive and possibly synergistic, in guiding the human brain towards a 'healed state'. We are learning the potential of the human brain's adaptability after insult and healing. While studies show there are potential pathways to not only treat the brain and try to remove it from a pathological path, but support it or protect it from inevitable outcomes, for instance, problematic parenchymal scarring. We are still trying to understand what the brain can do with the right intervention, at the right time, and in the right clinical context after considering all factors which contribute to healing and finding the optimum balance for maintained health and functionality, like the undisturbed nervous system does.

In: Movement 2019: Brain, Body and Cognition
Editors: Joav Merrick and Gerry Leisman
ISBN: 978-1-53616-714-6

Chapter 10

ENHANCING PELVIC MOBILITY AND SCHEMA IN DANCE PERFORMANCE WITH DYNAMIC NEURO-COGNITIVE IMAGERY

Eric Franklin* and Amit Abraham
The International Institute for Franklin Method, Zurich, Switzerland;
Emory University School of Medicine, Department of Medicine,
Division of General Medicine and Geriatrics, Atlanta, Georgia, United States of America;
and Department of Kinesiology, College of Education, University of Georgia,
Athens, Georgia, United States of America

Keywords: pelvis, mobility, schema, mental imagery

Objective: Successful dance performance requires high levels of pelvic awareness, control, and range-of-motion, which are all only available with highly accurate body schema: the dynamic, mental representation (i.e., picture) of the body, and its parts, in relation to the surrounding environment. Mental imagery (i.e., the creation and use of mental images and metaphors) is relevant for addressing and enhancing body, and specifically pelvic, mobility and schema. Dynamic Neuro-Cognitive Imagery (DNI) is an evidence-based, codified approach to mental imagery. DNI is designed for movement and postural retraining in both dancers and non-dancers. This approach focuses on the embodiment of human anatomy and correct biomechanics through mental imagery in order to improve body-related self-awareness as well as motor and non-motor aspects of performance. As such, DNI shows promise as a training approach for facilitating and enhancing pelvic mobility and schema.

* Corresponding Author's Email: franklinmethod@me.com.

Method: In this workshop, participants will 1) Learn how DNI training can address and enhance pelvic mobility and schema; 2) Understand key principles of mental imagery and DNI for enhancing bodily-related self-awareness; 3) Use DNI techniques and tools of anatomical and metaphorical imagery to experience and embody pelvic functional mobility with correct biomechanics; and 4) Learn specific DNI exercises for improving pelvic alignment and stability throughout performance of dance movements, such as plié and jumps.

At the end of the workshop, time will be devoted to questions and answers.

In: Movement 2019: Brain, Body and Cognition
Editors: Joav Merrick and Gerry Leisman
ISBN: 978-1-53616-714-6

Chapter 11

A NOVEL OUTLOOK ON CORRELATION BETWEEN ACUTE AND CHRONIC INFLAMMATORY STATES: 21 CASE STUDIES OF NEUROLOGICAL/PSYCHIATRIC/ MUSCULOSKELETAL DISORDERS

Seema Mahesh[1,*], Vitalie Vacaras[2], Viraj Shah[3], Elena Serzhantova[4], Nadezhda Kubasheva[5], Dmitriy Chabanov[4], Dionysios Tsintzas[6], Atul Jaggi[7], Mahesh Mallappa[1] and George Vithoulkas[8]

[1]Centre For Classical Homeopathy Bangalore, India,
[2]University of Medicine and Pharmacy, Iuliu Hatieganu, Cluj-Napoca, Romania,
[3]Shah Homeopathic Clinic, Ahmedabad, India,
[4]Novosibirsk Center of Homeopathy, Novosibirsk, Russia,
[5]Clinic of Nadezhda Kubasheva, Moscow, Russia,
[6]Rehabilitation Center of General Hospital of Aitoloakarnania, Greece,
[7]H3 Centre of Classical Homeopathy, Nashik, India
[8]University of the Aegean International Academy of Classical Homeopathy, Alonissos, Greece

Keywords: acute inflammation, sub-acute inflammation, chronic inflammation, suppression of fevers

Objective: To investigate if there is a correlation between the acute and chronic inflammatory states.

---

[*] Corresponding Author's Email: bhatseema@hotmail.com.

Method: Neurological/musculoskeletal and psychiatric cases, resulting in limitation in movement and normal sociocultural life, being treated with individualized classical homeopathy, showing significant improvement in their chronic pathology were investigated to see if acute inflammations or past suppressed diseases developed as the chronic states improved.

Result: The retrospective graphical analysis of 21 cases, truly improving, exhibit a specific pattern - as the chronic complaint improves the past acute inflammations, fevers and old suppressed simpler inflammatory states start re-appearing and increasing in frequency and that after these acute episodes (treated homeopathically) the improvement in the main complaint is even more apparent. In cases where the re-appearing acute inflammatory disease was treated with conventional drugs, the chronic complaint relapsed.

Conclusion: Chronic neurological/musculoskeletal/psychiatric pathologies improving under individualized classical homeopathic therapy exhibited development of acute diseases and return of suppressed simpler pathologies (e.g., skin eruptions) as the chronic disease improved. This may indicate an immunological shift from low grade chronic inflammation to efficient acute inflammation in the organism. Further studies may establish the connection between handling of the acute diseases with drugs and its role in development of chronic sub-inflammatory states.

In: Movement 2019: Brain, Body and Cognition
Editors: Joav Merrick and Gerry Leisman
ISBN: 978-1-53616-714-6

Chapter 12

EXPLORING MUSCULOSKELETAL REDUNDANCY USING NULL SPACE PROJECTION

***Dimitar Stanev*[1], *Constantinos Koutsojannis*[2,*]**
***and Konstantinos Moustakas*[1]**

[1]Department of Electrical and Computer Engineering, University of Patras, Patra
[2]Department of Physiotherapy Technological Educational Institute of Western Greece, Messolonghi, Greece

Keywords: null space solutions, musculoskeletal systems

Objective: This work aims to develop methods for handling redundancy in the process of modelling, simulation and analysis of musculoskeletal systems.

Method: It is an undeniable fact that musculoskeletal systems are inherently redundant, i.e., there are more degrees of freedom than those required to perform certain tasks and each degree of freedom is actuated by multiple muscles. This over-availability poses numerous challenges in the process of modelling and simulation that can negatively affect the validity of the models and the obtained results, rendering their application frequently inappropriate for clinical practice. The projection from a low- to a high-dimensional space is not uniquely defined (e.g., determine the muscle forces given a set of joint space generalized forces), however, this indeterminacy can be captured mathematically using the notion of null space.

Result: The main contribution of this work, is the calculation of the feasible muscle forces that satisfy both the action and the physiological muscle constraints. Its importance

---

[*] Corresponding Author's Email: ckoutsog@teiwest.gr; ckoutsog1@gmail.com.

is demonstrated in the context of joint reaction analysis, where we show that misinterpretation of the results is possible if the null space solutions are ignored.

Conclusion: Null space solutions, although typically ignored in musculoskeletal system simulations, offer deep insights and provide a much broader framework for modelling a wide range of phenomena arising in clinical practice, such as rigidity and slack muscle disorder.

In: Movement 2019: Brain, Body and Cognition
Editors: Joav Merrick and Gerry Leisman
ISBN: 978-1-53616-714-6

Chapter 13

A LEARNING PROCESS WHICH IMPROVES SYMMETRY IN THE DYNAMICS OF THORACIC ROTATION: A NEW HOPE FOR THE AMELIORATION OF MOTOR DISABILITIES

Vinicius Monteiro Diederichs*
Visiting kinesiologist, John Cabot University, Rome, Italy

Keywords: motor impairment, size scaling, neuromodulation, serial reaction time, motor sequence learning

Intersegmental joint dynamics permit generalisability, that is, a combination of joints to achieve maximum reachable amplitude of one degree of freedom. Generalisability contains the information to configure all lesser amplitudes using many degrees. This paper describes a treatment to reduce asymmetry in thoracic rotation, which appears to cause motor disabilities and pain. It proposes a learning process to recalibrate the neuromuscular system. This treatment is based on classical conditioning in which actors receive instructions to control a specific coordinate of the dominant hand — the conditioned stimulus (CS) — to be paired with an elastic force — the unconditioned stimulus (US). This pairing of CS with US generates a sequence of events, the conditioned response (CR). To facilitate control, the hand first reaches the target position constraining the overall degrees of freedom to only one. This reduces the burden on the CNS to deal with the indeterminacy of limb lengths, the regulation of joint rotation, and the combination of multiple joints for performing the motor task. From supine, the pelvis

* Corresponding Author's Email: info@monteirodiederichs.com.

is stabilized using specially designed equipment allowing one hand to ensure stability of the trunk while the other hand maintains its maximum reachable amplitude. The dynamics of this CR generates coupling, comparable to the dynamics described for coupling of posture and gait.

Method: All right-handed. Group 1, thirteen participants with acute motor impairment received three treatments on different days. In experiment 2, twenty-six healthy participants were randomly assigned into two groups and asked to perform the experimental treatment with the dominant or non-dominant hand, respectively, for comparison. Seven variables were measured: four range of motion (ROM), two perceived effort, and one pain.

Result: In Group 1, improvement in neck ROM was significant ($df = 62$, $t = -4.23$, $p < 0.01$). In Group 2 and 3, improvement in trunk symmetry was significant ($df = 18$, $t = -7.05$, $p < 0.01$). In conclusion, the treatment is able to trigger a mechanism that detects when a variable reaches a critical value and initiates a transition from the dynamics of the action system and task constraints. The after effect is highlighted as a neuromodulation impacting muscle tone with long lasting amelioration in motor disabilities and pain.

In: Movement 2019: Brain, Body and Cognition
Editors: Joav Merrick and Gerry Leisman
ISBN: 978-1-53616-714-6

Chapter 14

"AS TIME GOES BY": MOVEMENT LITERACY AND THE REPRESENTATION OF TIME IN CHILDREN'S SELF-GENERATED GRAPHICAL MOVEMENT REPRESENTATIONS

Shlomit Ofer*[*] *and Billie Eilam
Kibbutzim College of Education, Technology and the Arts,
Tel-Aviv, Israel and University of Haifa, Haifa, Israel

Keywords: movement literacy, self-generated representations, time representation

Objective: The presented study was designed to detect the ways and solutions to the challenge of self-generating graphical representations for dynamic movement events. Considering that time is an integral component of any movement experience, the representation of time is especially demanding due to its mostly unnoticed properties, and due to the various means of its culture modes of representation, some of them being indirect.

Method: The research was conducted as a part of a wider study focusing on movement literacy – the connections between an actual physical movement and its conceptual and graphical symbolic representations, and within the framework of qualitative research. 11 dance-movement sequences of three complexity levels regarding time-movement relations, were presented to the participants. The complexity levels included: a single movement event within each time-unit, synchronous movement events within each time-unit, and asynchronous movement events across time-units. The

[*] Corresponding Author's Email: shlomit.ofer@gmail.com.

participants, 16 4th grade girls (9-10 years old), learned these short movement sequences, trained in their conceptualization (recognizing, labelling) and performance, and developed graphical symbolic representations for them. These symbols in turn were deciphered and performed by a "blind decipherer", a girl who participated in learning and training phases, but was not exposed to the demonstrated sequence, and did not take part in the development process. The "developers" improved their representations in light of observed feedback perceived from the decipherer's performance until it matched the original movement sequence. The collected data included the graphical movement representation ("scrips") and the transcriptions of discussions among participants, which exposed some of their considerations while developing and deciphering the movement symbols.

Result: In both the single and synchronous cases, distinct events were utilized for representing time units, using a variety of cultural conventions to represent chronological progression. In the asynchronous case, participants brought forward in their representations either the movement events or the time units.

Conclusion: The various ways of representation generated by the participants exposed some of their understanding of the component of time as a part of the whole movement. It has also shed light on cultural means and conventions available for children this age for the representation of time, as well as some novel ways for graphical representation.

In: Movement 2019: Brain, Body and Cognition
Editors: Joav Merrick and Gerry Leisman
ISBN: 978-1-53616-714-6

Chapter 15

INTELLIGENT BRAIN-MACHINE INTERFACE FOR NEURO-FEEDBACK REDUCES ABNORMAL WORK ANXIETY

Constantinos Koutsojannis**, *Stavroula Kalantzi* and *Panagiotis Kostopoulos

Health Physics and Computational Intelligence Lab, Department of Physiotherapy
Technological Educational Institute of Western Greece, Aigion and FOCUSLab,
Piraeus, Greece

Keywords: neurofeedback, work anxiety, brain machine interface, artificial intelligence

Objective: EEG biofeedback, is an educational method that helps a trainee to control brain activity and is related to calmness and anxiety, as wave frequency has been positively correlated with cognitive performance in all ages, both in healthy individuals and in those with neurological conditions. The present research aimed to examine whether the training of subjects combined with a novel brain-machine interface developed using artificial intelligence (AI) approaches, for individual feedback adaptation, could improve the reduction of abnormal working anxiety.

Method: The Symptom Checklist-90-Revised and patient's self-reports were used as objective measures of treatment efficacy. The sample were used in all phases of the experiment consists of 30 adults, professionals with demanding working hours, while ten participants as a control group attended for a year sessions with a psychologist. The rest

* Corresponding Author's Email: ckoutsog@teiwest.gr.

following the same neuro-reactivity method with half of them using novel brain-machine interface, participants trained to reduce their work anxiety.

Result: After 30 sessions of EEG biofeedback within only three months, reported a significant reduction in anxiety-related symptoms compared with the control group. Higher negative association of the increase of alpha waves produced with the self-observed anxiety in the subjects of the experimental groups compared the control group. Additionally positive correlation of the growth of alpha waves produced with the general availability of subjects in the two experimental groups. Finally significantly higher percentage increase in alpha waves produced in the subjects of the second experimental group compared to the overall increase in alpha waves produced in the subjects of the first experimental group.

Conclusion: Intelligent brain-machine assisted neuro-feedback could be an effective treatment for anxiety disorder. The importance of research lies on the encouraging results of the neuro-reactive method using AI driven interface in stress management for future application of the method for training people with a strong sense of anxiety or anxiety disorders due to traumatic experiences such as mourning and injuries, preparing for participation in procedures of highly competitive procedures, such as examinations and sport meetings.

In: Movement 2019: Brain, Body and Cognition
Editors: Joav Merrick and Gerry Leisman
ISBN: 978-1-53616-714-6

Chapter 16

COGNITIVE STYLE DIFFERENCES IN VISUOSPATIAL PROCESSING ASSESSED USING A TASK INCORPORATING THE PERCEPTION OF BIOLOGICAL MOTION

Kaivo Thomson**, *Allan Voog and Anthony Watt
Estonian Sport Psychology Society, Estonia and College of Arts and Education,
Victoria University, Melbourne, Australia

Keywords: biological motion, cognitive style

Objective: The ability to perceive biological motion using point-light displays allows for differentiation in the evaluation of human movement. Cognitive style theory describes how an individual acquires knowledge and processes information. The main purpose of the present study was to examine possible cognitive style differences in visuospatial processing assessed using a task incorporating the perception of biological motion. It was hypothesized that those who field independent (FI) would make fewer errors than field dependent (FD) in determining similarities or differences between the point-light stimuli pair.

Method: A sample of 69 participants aged 18-27 years (M = 21.91, SD = 2.39) that included 33 males and 36 females completed the experiment. Cognitive style was assessed using the Group Embedded Figure Test. Perception of biological motion was evaluated using two different point-light stimuli developed from video images of a ballet dancer`s performance of a correct and incorrect turn in the fifth position. The point-light

* Corresponding Author's Email: kaivo.thomson@gmail.com.

stimulus format for exemplifying biological motion served as the basis of the task used to compare the visuospatial processing skills of the participants.

Result: A repeated measures MANOVA was conducted to evaluate differences between the FD and FI groups in relation to test occasion. A significant difference was found between the FD and FI groups in decision-making errors, $F(1,67) = 19.131$, $p < .001$. The effect size was small, ??² = .022. The within subjects analysis indicated no significant differences across test occasions, $F(1,67) = 0.34$, $p = .124$. No significant difference was found, $t(67) = 0.64$, $p = .53$. Overall, the results clearly highlight that the FI group made less errors than the FD group at each test occasion. Results clearly supported the research hypothesis and revealed that individuals who were classified as field independent made significantly fewer errors at each test occasion.

Conclusion: The current findings showed that performance on the point-light discrimination task could be differentiated on the basis of preferred cognitive style. Furthermore, this study may contribute to a clearer understanding of the processes that individuals use to decode spatiotemporal information associated with perception of biological motion.

In: Movement 2019: Brain, Body and Cognition
Editors: Joav Merrick and Gerry Leisman
ISBN: 978-1-53616-714-6

Chapter 17

PSYCHOMETRIC PROPERTIES OF MOTOR AND COGNITIVE DUAL-TASK STUDIES WITH THE AIM OF DEVELOPING A TEST PROTOCOL FOR VESTIBULAR-IMPAIRED PATIENTS

Maya Danneels[1,*], Ruth Van Hecke[1], Hannah Keppler[1,2], Sofie Degeest[1], Dirk Cambier[1], Raymond van de Berg[3,4], Vincent Van Rompaey[5,6] and Leen Maes[1,2]

[1]Ghent University, Department of Rehabilitation Sciences, Ghent, Belgium,
[2]Ghent University Hospital, Department of Otorhinolaryngology, Ghent, Belgium,
[3]Maastricht University Medical Center, Department of Otorhinolaryngology and Head and Neck Surgery, Maastricht, Netherlands,
[4]Faculty of Physics, Tomsk State Research University, Tomsk, Russia,
[5]Department of Otorhinolaryngology and Head and Neck Surgery, Antwerp University Hospital, Edegem, Belgium
[6]Faculty of Medicine and Health Sciences, University of Antwerp, Antwerp, Belgium

Keywords: vestibular impairment, cognition, posture, systematic review, test protocol

Objective: Patients with vestibular disorders often suffer from cognitive impairment. On the one hand, this can be attributed to extensive vestibular projections throughout the cerebral cortex and sub cortex. On the other hand, increased cognitive-motor interference (CMI) might present in persons with vestibular disorders. CMI can be assessed by performing dual-tasks, however, literature on this topic is scarce in the vestibular

* Corresponding Author's Email: maya.danneels@ugent.be.

population. Therefore, before implementing a dual-task protocol in these patients, a systematic review on the psychometric properties of dual-tasks in a variety of populations was conducted.

Method: The systematic review was conducted in accordance with the Preferred Reporting Items for Systematic Reviews and Meta-Analyses (PRISMA) guidelines. An extensive literature search was run on MEDLINE, Embase, and Cochrane Databases. Their eligibility was assessed by two independent researchers, and their methodological quality was evaluated using the Consensus-based Standards for the selection of health Measurement Instruments (COSMIN).

Result: Thirty-three studies were included in the current review. The cognitive and motor tasks varied in terms of test-retest reliability, and some were valid in persons with Parkinson's disease, multiple sclerosis, dementia, stroke, or elderly.

Conclusion: In order to maximally evoke CMI in the vestibular population, both static and dynamic motor tasks should be performed while challenging the vestibular cognitive domains (visuospatial abilities, memory, attention, executive function, and processing speed). Out of the large amount of cognitive tasks employed in dual-task studies, multiple tests can be suggested for executive function and processing speed. For the cognitive domain memory, only the Backward Digit Span Test can be recommended. As attention is required in every dual-task protocol, this component cannot be challenged separately. Finally, visuospatial cognition was not assessed in a reliable or valid way in the included dual-task paradigms. The use of dual-tasks will give a more accurate and daily life representation of cognitive and motor deficiencies and their interaction in the vestibular-impaired population.

In: Movement 2019: Brain, Body and Cognition
Editors: Joav Merrick and Gerry Leisman

ISBN: 978-1-53616-714-6

Chapter 18

INTENSIVE NEUROREHABILITATIVE TRAINING AIMING TOWARDS TRIGGERING BRAIN REMAPPING AND REMYELINATION IN MULTIPLE SCLEROSIS PATIENTS

Triantafillia Reziti*[*]*, Fotini Bazoti, Polixeni Kontou and Irene Alexopoulou
Feldenkrais Praxis and Somatics Center, Perama, Greece

Keywords: Feldenkrais method, multiple sclerosis, self-Image development, remyelination, embodied integration

Objective: Development of a strategy in order to promote the embodiment of changeability. This involves connecting and responding as a moving self that changes and synthesizes different body part movements simultaneously, while editing the parameters of intention, time, space, direction, speed and in/out switch of attention.

Method: Trial of a patient group (PG) with Remitting Relapsing (RR) and Primary Progressive (PP) Multiple Sclerosis (MS) (n = 9), suffering from sensory and motor deficits, following an intensive protocol (four days, four hours per day) based on the Feldenkrais Method (FM) and Embodied Life™ School (ELS) standing gestures. The PG was evaluated before and after the intensive training protocol comprising of a) Awareness Through Movement (ATM) classes, b) Embodied Standing Archetypical Gestures and patterns with universal context (reaching, showing with index finger, intention to go), c) mirroring techniques through choreography, d) breathing experiments (exhalation, inhalation, activation of the swallowing reflex), e) analysis of walking patterns (motor control, connection between consciousness and gravity), f) employment of darkness for developing spatial orientation and positional sense, g) motor imagery using body parts, like tongue, that have large representation in the primary motor cortex. The evaluation of PG was performed by: a) Berg Scale, to assess balance and movement efficiency, b) Mood Score, to assess self-estimate, and c) the developed Protocol Scale, to assess neuroplasticity efficiency.

Result: The PG that received the developed intensive protocol expressed a positive experience of self-image and improved motor imagery ability. After only four days, individuals in the PG composed movement patterns that significantly reinforced their embodied activities (balance, walking, spatial orientation, velocity and basic functional organization). The developed protocol enhanced the inner/outer world interaction by reducing cognitive obstacles, thus providing better spatial adaptation in the continuous environmental changes without interruptions. This drove the formation of a buffering condition to neurosomatic movement compilation.

Conclusion: The developed intensive FM/ELS-based training protocol was effective in triggering brain remapping and remyelination in MS patients.

In: Movement 2019: Brain, Body and Cognition
Editors: Joav Merrick and Gerry Leisman
ISBN: 978-1-53616-714-6

Chapter 19

MOVEMENT-BASED PREVENTION OF EMOTIONAL-BEHAVIORAL DISORDERS IN ELEMENTARY SCHOOL CHILDREN

Anne Menke*[*] *and Satyam Antonio Schramm
Department of inclusive Education-Emotional and Behavioral Disorders (EBD),
University of Potsdam, Potsdam, Germany

Keywords: cognitive skills, self-regulation, behavioral problems, executive functions

Objective: Only one quarter of all children meet the demands of 60 minutes daily movement activity as recommended by the World Health Organization. That is, even though promoting physical activity at a young age can contribute towards prevention of attention deficit/hyperactivity disorder and can lead to greater cognitive and academic achievements. Furthermore, an increasingly important aspect is the interaction between physiological and mental development. The positive relation between general fitness, cognitive functioning and psychological wellbeing has been found in several studies. The current study presents first implications of the feasibility and efficacy of a movement-based prevention program for emotional behavioral disorders in elementary school children. This Program ("Strong muscles, alert mind") took place in a group setting (12 weekly sessions for 3-4 months) and each session contains three elements (proprioceptive training; strength, endurance and coordination; strength endurance and variability).

[*] Corresponding Author's Email: annemenke@uni-potsdam.de.

Method: In our study children (n = 60) aged 4-10 years, participated in the program "Strong muscles, alert mind". Outcomes evaluated the participant's coordinative skills (DCDQ) as well as their social emotional problem behavior (SDQ) rated by their parents in a pre-post design. Dependent T-Tests were used to test for significant change from pre to post, followed up by calculation of effect sizes.

Result: Statistical analyses reveal positive effects not only for the outcome of coordinative skills but also for social-emotional problem behavior. Significant changes can been found on all subscales of the DCDQ ($p < .01$) as well as on three of the four subscales of the SDQ ($p > .05$).

Conclusion: The study's results are promising and go along with the latest research stating that movement effects cognitive skills and brain activities in a positive way (Bös & Obst, 2000) Nevertheless, there are several limitations that need to be addressed in future research. One of the main aspects is the missing control group and therefore the need of a follow-up study in a randomized control group design with a multiple informant approach.

In: Movement 2019: Brain, Body and Cognition
Editors: Joav Merrick and Gerry Leisman
ISBN: 978-1-53616-714-6

Chapter 20

RECONNECTING TO THE SELF: WORKING WITH REFUGEES USING DANCE THERAPY AND LABAN/BARTENIEFF MOVEMENT STUDIES

Rena Milgrom*
Open field, NGO, Czech Republic

Keywords: dance therapy, Laban, trauma, intervention

We are facing a global crisis, wherein dealing with the integration of refugees into our societies is becoming an increasingly difficult issue. Dance Movement Therapy and Laban/Bartenieff Movement Studies can offer possibilities for working with groups or individuals facing this situation.

This workshop presents interventions developed during my work with Afghan refugees in the Czech Republic.

Regardless of nationality, I would like to point out issues faced by refugees or immigrants finding themselves in a different cultural and social environment.

From the viewpoint of information transfer, there is a loop going from perception, sensing, feeling, through to emotional states and on to beliefs, judgments, inner dialogues and the processing bodily experience. A loop that moves from the brainstem to the limbic system to the neocortex.

Common emotionally charged features of a refugee's experience can be oscillating mixtures of reflexive defenses, stress, fear, sadness, loss of autonomy, a broken sense of safety and connection, heightened alert and struggle for hope. These characteristics are

---

* Corresponding Author's Email: soma@somatika.cz.

mirrored in the body's posture, gestures, in effort flow, attention, in relationships as well as in a spatial disorientation.

As a tool for reconnecting the Self, I would like to propose re-patterning using several principles, concerning:

- deepening contact with the body (breath support, weight sensing, flow)
- connecting with others (interaction, adaptation)
- orientation (Space Harmony using Laban movement scales)
- establishing personal space (Kinesphere)
- empowerment (weight activation)
- movement meditation with imagery
- awareness of personal needs and boundaries

In: Movement 2019: Brain, Body and Cognition
Editors: Joav Merrick and Gerry Leisman
ISBN: 978-1-53616-714-6

Chapter 21

WHY DO ONLY SOME ARM RECOVER AFTER STROKE? A STORY OF MIRROR NEURONS IN BA44

Deanna Anderlini [*] ***and Guy Wallis***
Centre for Sensorimotor Performance, HMNS, University of Queensland, Brisbane and Neurology Department, Royal Brisbane and Women's Hospital, Brisbane, Australia

Keywords: stroke, aphasia, mirror neurons, motor recovery

Objective: Cortex and cortical function can be regarded as modular, with separate areas involved in processing sensory information and the initiation of motor movement. The blood supply supporting cortex is, like the brain itself, modular. Disruption in blood flow leads to a multiplicity of dysfunctions like the frequent co-occurrence of aphasia and right upper limb hemiparesis.

A review of the stroke literature reveals a correlation between Broca's aphasia and upper limb motor recovery. The precise reason for the correlation is not known but this paper proposes one. Our argument is that speech impairment is indicative of damage to Brodmann area BA45 but that the motor deficits are due to damage to the proximal, but functionally discrete area BA44. BA44 is a multisensory area. But experiments on tone-deaf or stutters, radiological tools like fMRI and DWI, studies of the neuro-ontogeny and development in babies, findings of genetic, epigenetic and embryology, all point to BA44 playing a central role in visuo-motor integration.

Method: We have analyzed data from 3780 stroke patients. Two groups were compared: patients with aphasia and right upper limb motor deficit and patients with just motor deficit without aphasia.

[*] Corresponding Author's Email: Deanna.anderlini@uqconnect.edu.au.

Result: Patients with Broca's aphasia show slower and poorer motor recovery of right arm compared to non-aphasic ones.

Conclusion: Intact BA44 offers a source of unimpaired input to a damaged motor system from primary visual pathways and the corticobasal-ganglia thalamic loop. However, when damaged, the motor system is starved of one important source of signal for effective retraining and recovery.

In: Movement 2019: Brain, Body and Cognition
Editors: Joav Merrick and Gerry Leisman
ISBN: 978-1-53616-714-6

Chapter 22

EFFICACY OF USING MOVEMENT INTERVENTION IN ADULTS LIVING WITH CEREBRAL PALSY: A QUALITATIVE STUDY

Nina Martin[*]
School for Classical and Contemporary Dance, Texas Christian University,
Fort Worth, Texas, United States of America

Keywords: movement, cerebral palsy, dance, therapy, adults

Objective: To qualify a movement intervention, ReWire Movement Method (RMM), for adults living with Cerebral Palsy (CP) who have reduced programs once aging out of the USA system after 18 years of age. To assess, qualitatively, the effects of the RMM dance-derived activity on quality of life for participants who are all dancers in a mixed-ability dance company and living with CP. Participants reported improvements from the practice of RMM. The study focused on three participants, all adults with spastic cerebral palsy (no cognitive impairment) using wheelchairs and crutches (GMFCS grade 1/2/3).

Method: Qualitative research methods investigated the efficacy of RMM for inducing short-term cessation/reduction of spasms, increased range of motion, heel to floor proximity, improved gait and posture, improved fine motor control and sense of well-being. Interviews provided first-person data before and after RMM practice.

[*] Corresponding Author's Email: nina.martin@tcu.edu.

Result: Multiple interviews conducted with the participants before and after the movement intervention substantiated anecdotes suggesting RMM improves quality of life while reducing anxiety, depression, and fatigue.

Conclusion: Results suggest that ReWire Movement Method, as a movement practice, provides significant (short term) improvement in quality of life for participants. Further studies of RMM are warranted based on this study. The practice has potential to impact a broad population and should be established considering the ease of implementation which does not require special apparatus, but only a trained facilitator. Once participants learn RMM they can practice it on their own.

In: Movement 2019: Brain, Body and Cognition
Editors: Joav Merrick and Gerry Leisman
ISBN: 978-1-53616-714-6

Chapter 23

THE CHALLENGE OF COORDINATION: A PRACTICAL WORKSHOP

Lilach Shalit [*] ***and Tali Ronen***
School for the Arts of Dance, Faculty of the Arts, Kibbutzim College of Education, Technology and the Arts, Tel-Aviv, Israel;
Department of Dance, Orot Israel Academic College of Education, Elkana;
and School of Physical Education and Sports Ohalo College, Katzrin, Israel

Keywords: coordination, motor skills, cognitive abilities

The proposed workshop engages in physical practice targeted at improving cognitive skills. The structure of the physical exercises will be explained and then performed by the participants in purpose of experiencing the complexity and the challenges of the motor execution.

The leading principles of the workshop are based on the Sapir Method, which focuses on the improvement of learning skills, cognitive abilities and motor abilities. The method has been applied among various populations (learning disabilities, age-related effects, etc.) within a wide range of age group (children, adolescences, elderly).

The Sapir Method is anchored in Eshkol-Wachman Movement Notation (EWMN) a numerical-graphic language that describes human movement in space and time. EWMN is used for analyzing different movement disciplines, creating and teaching. EWMN is capable of representing complex movement events, including movements that are performed in simultaneous form. The structure of the notation exposes different movement layers which can be performed separately or simultaneously in varied

[*] Corresponding Author's Email: lilach.g.s@gmail.com.

combinations. This way of thinking and analyzing expresses many coordination possibilities.

The motor practice will focus on the concept of coordination, which includes cognitive processes and motor execution. Coordination is an interaction between each of the central nervous system, skeletal and muscles. Several of coordination structures will be demonstrated – from simple to complex, each of which demands a high level of attention allocation. One of the basic principles that will be emphasized is the creation of new and original movement sequences, since automated movements do not lead to further development of coordination.

The workshop is designed for all populations (not only for dancers), thinking that this kind of practice can be meaningful and important to all, therefore no prior knowledge is required. The center of this workshop is the personal experience which is based on the physical practice that could be both frustrating and joyful. Coordinative practice can be used as a bridge between movement and cognition. This is its mystery and its challenge.

In: Movement 2019: Brain, Body and Cognition
Editors: Joav Merrick and Gerry Leisman
ISBN: 978-1-53616-714-6

Chapter 24

DEVELOPMENT OF MULTIPLE INTELLIGENCE

Patricia Elizabeth Torres Villanueva*
Research Center of Fine Arts, Universidad de Colima, Colima, Mexico

Keywords: brain enhancement, neuroplasticity, cognitive development

Objective: Mental and emotional health improvement by interacting and sharing self-development and proprioceptive experience.

Method: Ancient knowledge in contemporary society teaching circle dance to develop and exchange cognitive abilities through observation, concentration and rhythmical repetition

Result: Physical and emotional awareness and conscious awakening towards our ecological and intangible spiritual surroundings

Conclusion: Effective implementation as pedagogical and therapeutic tools in schools, universities, hospitals, rehabilitation centers and social communities.

* Corresponding Author's Email: tezcaxx1@yahoo.com.

In: Movement 2019: Brain, Body and Cognition
Editors: Joav Merrick and Gerry Leisman
ISBN: 978-1-53616-714-6

Chapter 25

THE ANTICIPATORY MOVEMENT

Zdravko Radman*
Institute of Philosophy, Zagreb, Croatia

Keywords: philosophy of mind, extended mind, action, movement, anticipation

One of the influential theories in the contemporary philosophy of mind and consciousness is that of the 'extended mind'. In this talk I want to expound the idea that mind is 'extended' also temporally; that anticipation is one of the fundamental features of the mind. The term 'aheadness' will be coined and used to account for the prospective nature of our mentality, that is for the capacity of the mental to shape intentional contents of which there are not yet evidence in the senses. All of our mentality is strongly future-oriented – providing the scenarios of what to the cognitive organism appears to be most plausibly the case in the world. This will be exemplified on the case of movement and, more generally, on the generation of action. Because an agent has to be prepared for the 'next step' he undertakes (otherwise his reactions will be late), he projects 'wild stories', that is, hypothesizes about the forthcoming by drawing on the past experience. Movement, in that sense, never starts blind or uninstructed for it happens in environmental situations already patterned by anticipations that turn mere surroundings into an arena of possible movements and behavior.

* Corresponding Author's Email: radman@ifzg.hr.

In: Movement 2019: Brain, Body and Cognition
Editors: Joav Merrick and Gerry Leisman
ISBN: 978-1-53616-714-6

Chapter 26

PERCEPTION OF TIME: VARIABILITY IN DURATION DISCRIMINATION OF SHORT AUDIO STIMULI

Ivan Šerbetar*
Department of Kinesiology, Faculty of Teacher Education,
University of Zagreb, Zagreb, Croatia

Keywords: time perception, discrimination threshold

Objective: Time estimation is crucially important for almost every human behavior. Brain uses knowledge of time's passage to anticipate sensory events and to arrange motor and cognitive actions accordingly. One of the key questions in the field is how the nervous system encodes the events' duration in the milliseconds range, which is particularly relevant for instant perception and the actions produced and related to these events.

The goal of current study is to test discrimination of audio intervals in milliseconds range.

Method: Ten subjects participated in the pilot experiment. Psychoacoustics - the MATLAB toolbox for auditory threshold estimation by Soranzo and Grassi (2014) was used to establish the discrimination threshold as the minimum detectable difference in duration between the two tones. The duration of stimuli was selected in real time during the experiment and each trial depended on the previous response. The standard interval was set to 200 msec and the initial comparison interval was 100 msec; step size or value of reduction/increment of the tone duration was fixed. Each subject performed 5 blocks, and each block contained 50 trials.

* Corresponding Author's Email: ivan.serbetar@ufzg.hr.

Result: The pilot data were significantly skewed and kurtotic for the first block of trials, but not for the others. The overall mean was M = 39.79 (msec) without clear pattern of reduction or increment from the first to the last block. The SD was highest (20.35) in the first block with the lowering tendency toward the end of trials (10.92). Subjects' individual thresholds ranged from 23.51 to 56.03 msec but individual variability (SD) was lower than overall. Repeated measures ANOVA on block discrimination thresholds showed no significant effect of time.

Conclusion: Within variability in duration discrimination was lower than between which points to relatively stable human ability. However, some technical issues like step size and number of steps should be considered to eventually lower the variability between blocks, mostly caused by attention and/or memory lapses.

In: Movement 2019: Brain, Body and Cognition
Editors: Joav Merrick and Gerry Leisman
ISBN: 978-1-53616-714-6

Chapter 27

LABAN MOVEMENT ANALYSIS AS A FRAMEWORK FOR COGNITION OF MOVEMENT IN RESEARCH AND INTERVENTION

Rachelle Palnick Tsachor*
School of Theatre and Music, College of Architecture, Design, and the Arts,
The University of Illinois at Chicago, Chicago, Illinois, United States of America

Keywords: Laban Movement, movement, intervention

Movement cognition can be enhanced by a comprehensive vocabulary inclusive of qualitative, first-person experiential, and objective, observable aspects of movement. We present diverse uses of Laban Movement Analysis to facilitate cognition of movement for research and intervention. Discussion explores a continuum from objective observation of movement to individual intervention.

Body movement is ever more recognized as integral to both mental and physical health. Movement therapies such as Dance/Movement Therapy (DMT), Physical Therapy (PT), Somatic Movement Education & Therapy (ISMETA) use body movement as a primary tool to support and improve functional and psycho-social well-being through physical, somatic, creative and expressive movement processes.

Identifying which aspects of movement are responsible for the effects of different movements is challenging because movement is a complex phenomenon, with multiple, overlapping variables, uniquely presenting in each individual and circumstance. A sophisticated taxonomy can facilitate recognition of movement's content and context.

---

* Corresponding Author's Email: rtsachor@uic.edu.

Laban Movement Analysis (LMA) is an internationally-accepted, comprehensive system identifying movement components, with a strong record of inter-observer reliability. LMA (and systems emerging from it) is used in research for its ability to support cognition of qualitative aspects of movement as they shift or remain constant over time. LMA facilitates movement pattern recognition at macro and micro-analysis levels. It is valued for accuracy observing subtle and momentary changes, functional and expressive aspects of movement, and language-based descriptors which align with how people think about movement. LMA-based systems provide a taxonomy supportive of both research and intervention.

This symposium explores diverse applications of LMA-based systems in research, theory, psycho-therapy, physical therapy and personal development.

In: Movement 2019: Brain, Body and Cognition
Editors: Joav Merrick and Gerry Leisman
ISBN: 978-1-53616-714-6

Chapter 28

OVERVIEW: HOW THE LABAN/BARTENIEFF MOVEMENT TAXONOMY SUPPORTS COGNITION OF MOVEMENT IN RESEARCH AND INTERVENTION

Rachelle Palnick Tsachor*
School of Theatre and Music, College of Architecture, Design, and the Arts,
The University of Illinois at Chicago, Chicago, Illinois, United States of America

Keywords: Laban movement analysis, Laban/Bartenieff movement system, movement research

This symposium explores diverse applications of Laban Movement Analysis (LMA)-based systems in research, theory, psycho-therapy, physical therapy and personal development. Body movement is ever more recognized as integral to both mental and physical health. Movement therapies such as Dance/Movement Therapy (DMT), Physical Therapy (PT), and Somatic Movement Education & Therapy (ISMETA) use body movement as a primary tool to support and improve functional and psycho-social well-being through physical, somatic, creative and expressive movement processes. DMT is a well-established therapy with clinical evidence of effectiveness. Somatic Movement Education and Therapy is broadening field with evidence of effectiveness.

LMA has a long research history studying: emotions; clinical assessment; polio; rehabilitation, cross-cultural studies, personality, efficiency, and style. Its versatility is evident from its use in diverse movement research: animal behavior; autism; stroke;

* Corresponding Author's Email: rtsachor@uic.edu.

infant development; gaming/robotics/animation; brain mechanisms; musical expression; emotion and decision-making.

Identifying which aspects of movement are responsible for the effects of different movements is challenging because movement is complex, with multiple, overlapping variables. A sophisticated descriptive taxonomy can facilitate recognition of movement's content and context. LMA is an international, comprehensive system for naming movement components, with a strong record of inter-observer reliability. LMA-based systems' language-based taxonomy bridge from research to intervention because they identify qualitative and quantitative aspects of movement as they shift over time so we can recognize patterns at both the macro and micro-analysis levels. Valued for its accuracy in observing subtle and momentary motor changes, functional and expressive aspects of movement, its language-based descriptors align with how people think about movement (e.g., 'touch lightly' vs. kinematics).

In: Movement 2019: Brain, Body and Cognition
Editors: Joav Merrick and Gerry Leisman
ISBN: 978-1-53616-714-6

Chapter 29

HOW THE BODY SPEAKS: THE NONVERBAL CUES OF TRAUMA BETWEEN MOTHER AND TODDLER

Suzi Tortora*
Dancing Dialogue and Memorial Sloan-Kettering Cancer Center (MSKCC),
New York, United States of America

Keywords: Laban nonverbal movement analysis, video-feedback intervention, maternal reflective functioning, dance/movement psychotherapy, PTSD

Objective: Video-feedback used to support mother-child attachment relationships is common practice in "at risk" infant mental health programs. This presentation introduces Laban Movement Analysis-based observational tools developed by the presenter, used to analyze interpersonal violence-exposed mothers with PTSD (IPV-PDSD) and their toddlers, using video excerpts from Dr. Schechter and colleague's research in Geneva, Switzerland and New York: Clinical Assisted Video-feedback Exposure Sessions (CAVES). Implications for treatment will be discussed.

Method: The parent-toddler dyad is analyzed using a nonverbal analysis tool: Dyadic Attachment-Based Nonverbal Communicative Expressions (D.A.N.C.E.), comprised of ten nonverbal movement analysis categories. D.A.N.C.E. analyzed three videotapes of the CAVES protocol: 1. feedback session of mother and clinician watching a video of the dyad playing; 2. dyad playing; 3. dyad with the clinician(s).

Result: D.A.N.C.E. provides insight into the quality of the mother-child attachment relationship. Increased maternal reflective functioning obtained through CAVES intervention corresponded to more fully coordinated embodied nonverbal behaviors rather than triggered traumatic-startle nonverbal reaction.

Conclusion: D.A.N.C.E. findings provide intervention strategies to support improved mother-child interactions with the “at risk” population. These strategies include increasing the mother’s ability to more accurately read her child’s nonverbal cues and increased awareness of the implications of her own nonverbal actions. Activities to support playful ways to improve social emotional engagement between the mother and toddler using movement and dance activities is also discussed.

In: Movement 2019: Brain, Body and Cognition
Editors: Joav Merrick and Gerry Leisman
ISBN: 978-1-53616-714-6

Chapter 30

MOVING OUT THE FEAR OF MOVEMENT: LABAN MOVEMENT ANALYSIS APPLICATIONS IN A POPULATION WITH KINESIOPHOBIA DUE TO CHRONIC PAIN CONDITIONS

Irena Paiuk*
School of Health Professions, Physical Therapy Department, Tel-Aviv University, Tel Aviv, Israel

Keywords: musculoskeletal pain, kinesiophobia, LBMS, movement

Pain is a clinical component of many chronic musculoskeletal conditions, and is defined as an unpleasant sensory and emotional experience associated with actual or potential tissue damage. The experience of pain is modulated through emotions and influences cognitions and behaviors. Positive emotional states generally reduce pain while both fear and anxiety augment it. In acute stages of injury common responses to pain are kinesiophobia (fear of movement) and fear avoidance behaviors, which are associated with poor pain management, higher risk of developing chronic musculoskeletal pain and disability. Although pain and kinesiophobia change one's functional and expressive aspects of movement, clinical assessment of kinesiophobia is performed mainly with questionnaires. A skilled framework for movement analysis and identification of movement characteristics of kinesiophobia seems to be lacking in clinical practice. Additionally, current clinical interventions for kinesiophobia use mainly a Cognitive Behavioral Approach (CBT) than movement based approach. According to Laban Movement Analysis (LMA), movement is a mutual psycho-physical dialogue, in which personal intents, sensation and emotions are influenced by movement and are reflected through movement in terms of Patterns of Body Organization, fluctuations in Breath Support, movement Phrasing, Efforts, Shape Qualities or Approach to Kinesphere.

The objectives of the study are 1. To recognize and define movement patterns of kinesiophobia through the lens of LMA, by analyzing case-studies from a clinical practice; 2. To analyze how LMA approach may facilitate a positive change in kinesiophobic patients and expand their movement potential.

Result: Movement manifestations of Kinesiophobia due to chronic musculoskeletal pain which were identified were Bound Flow, Enclosing and Condensing Shape Qualities, changes in Core-Distal connectivity and in Breathing Phrasing. LMA based interventions provoked a change in patients' body sensations and movement qualities, and consecutively promoted positive emotional changes, expanded patients' kinesphere and decreased pain. Diminished sense of fear of movement was reflected through a more affluent patients' movement qualities.

Conclusion: Identifying movement patterns of kinesiophobia due to chronic musculoskeletal pain using LMA may allow the therapist a route to facilitate the unfolding of patents' supportive movement choices to overcome the Kinesiophobia, and in such way to move the fear out.

In: Movement 2019: Brain, Body and Cognition
Editors: Joav Merrick and Gerry Leisman
ISBN: 978-1-53616-714-6

Chapter 31

MOVING OURSELVES: CREATIVE INTERVENTIONS FOR CHANGE USING THE LABAN/BARTENIEFF MOVEMENT SYSTEM

***Joanna Brotman*[*] *and Katharina Conradi*[†]**
Laban Bartenieff Institute of Movement Studies, Embodied Movement Institute,
Rutgers University, New Brunswick, New Jersey, United States of America
and University of the Arts Rotterdam, Academy of Dance and Theatre Amsterdam,
Embodied Movement Institute, Rotterdam, the Netherlands

Keywords: Laban Bartenieff, movement patterns, self-development, moving our selves

Objective: Moving our Selves is an intensive movement-cognition program integrating theoretical and experiential aspects of the Laban Bartenieff Movement System (LBMS) to support participant's ability to identify how our movement choices can hinder and support the changes we seek. The LBMS cognitive framework enables participants to perceive, experience, and describe their own personal movement patterns, and to become cognizant of how these patterns influence the way they negotiate boundaries and balance the needs of self and other. Participants then creatively design ways to shift their own patterns to support meaningful and productive change in their personal, artistic, and professional lives.

Method: In Moving Our Selves, the LBMS-based method we designed allows participants to gather information about their own movement choices and preferences through guided experiences. Each participant synthesizes their own movement research

[*] Corresponding Author's Email: Joanna13x@gmail.com.
[†] Corresponding Author's Email: katharinaconradi@gmail.com.

data by creating a personal movement profile using graphic representation process that reveals the relationship between the parts and the whole of their patterns as it illuminates their correlative and associative aspects between movement and its personal meaning. Through cognition of movement patterns, each person uncovers the unique phrasing with which they organize movement to function and express themselves in the world. Participants then play creatively with their graphic representation to formulate personal interventions as a way to try on new possibilities.

Result: Moving Our Selves works with the premise that while the patterns each of us have developed may serve us well at times, we may also wish to have more choices in how we operate. Using LBMS, participants learn to recognize when, where, how and why they tend to "get stuck" in certain movement/life habits. From there, they explore ways to literally move through their "stuck state" as they discover new movement options in various phrasing contexts.

Conclusion: The Laban Bartenieff Movement System provides a taxonomy of what, where, when, how and why movement patterns occur, offering us a map of possibilities to connect to exiled parts of ourselves and discover new aspects of who we are becoming. Through this process, participants work towards developing a self–sustaining practice in support of readiness and embodied engagement.

In: Movement 2019: Brain, Body and Cognition
Editors: Joav Merrick and Gerry Leisman
ISBN: 978-1-53616-714-6

Chapter 32

SPATIAL COGNITION IN MOVEMENT: THE BODY-SPACE RELATIONSHIP TO THE FORMS OF THE ICOSAHEDRON AND ITS DUAL, THE DODECAHEDRON

Karen A Studd*

Embodied Movement Institute, The Netherlands and Inspirees Institute, China

Keywords: Laban/Bartenieff Movement System, Laban's Space Harmony, spatial cognition in movement, geometric forms in movement

Objective: A unique aspect of Laban/Bartenieff Movement System (LBMS) as a movement approach to the whole person is its capacity to address both context and the container of the mover's body in spatial relationship to one's environment. Through its theory and practice of Space Harmony, the framework of a Body/Space Duality is a very real representation of the primary duality of Self/Other: Self as identified with the Body and Other as the Environment (everything that is not self). This paper lays out how conceiving of space in different geometric forms changes the movers' spatial emphasis.

Method: Laban's Space Harmony theory is a unique practice linking aspects of the practice of conceiving of movement through geometric forms with somatic practice when exploring the patterns and experience of human movement. LBMS is an early system for mapping our Kinespheric Space. From a somatic perspective, our perception of space exists on a continuum from our Inner Space to the Kinespheric reach Space to the General Space beyond our physical reach, and to the infinite space of the cosmos. This

* Corresponding Author's Email: krnstudd@gmail.com.

paper examines movement in two of the many geometric patterns our cognition explicates in the entirety of the Space in which we move, and how cognition and intention of a spatial form affect body movement in space.

Result/Conclusion: The Body-Space relationship differs when the mover conceives of moving in the Icosahedron and its dual, the Dodecahedron. Examples presented include movement in our primary patterns of movement (for example walking) in relationship to our body's architecture, our experience of inner Body Space and the Kinespheric reach Space. This presentation explores logical extension/expansion of the Laban/Bartenieff Movement System (LBMS) theory and practice.

In: Movement 2019: Brain, Body and Cognition
Editors: Joav Merrick and Gerry Leisman
ISBN: 978-1-53616-714-6

Chapter 33

Relationship between Humans and Their Acoustic Environments

Kaitlin Sly*
University of Victoria in Victoria, British Columbia, Canada

Keywords: infrasound, ultrasound, noise, anthropogenic noise, soundscape

Objective: Certain sound frequencies, specifically those in the infrasound and ultrasound ranges, can have significant effects on human health and wellbeing. Infrasound and ultrasound frequencies are of particular concern because these are beyond the human hearing range, which suggests there may be an assumption amongst the general public that they do not have an effect on us.

Method: Through establishing and detecting where these frequencies are in the environment, this information can then be used to improve urban planning and design, so that we might more easily avoid their potentially adverse health effects. In order to do this, my project used new methods and instrumentation to measure and assess noise sources in public spaces, and synthesized these findings with studies in human auditory perception, specifically of noise hazards from infrasound and airborne ultrasound. More specifically, my project determined where anthropogenic sources of infrasonic and ultrasonic sounds were present in specific areas of Greater Victoria, through data collection and analysis (field recordings) over a 3-4 month period. Using the Earthworks M50 omnidirectional acoustic measurement device, which has an ultra-wide frequency response of 3Hz to 50kHz - wide enough to capture those frequencies in the infrasonic and ultrasonic spectrums, I investigated those locations that have previously been

* Corresponding Author's Email: kaitiesly@gmail.com.

reported to contain these frequencies, such as near highways. Through preliminary consultations with BC government representatives, I have discovered that the Ministry of Transportation and Infrastructure does not consider infrasound in their highway noise mitigation efforts, despite the fact that there is research suggesting these frequencies are present. According to T.G. Leighton, guidelines for ultrasound are also outdated, A-weighted, and are based on occupational exposures, which are different from public exposures.

Result: Results indicated that there are noteworthy public exposures to either infrasonic or ultrasonic (or both) sound frequencies in the locations where data was taken. The results from this project were disseminated in a creative work that was designed using Max/MSP and Jitter.

Conclusion: In investigating where infrasonic and ultrasonic sound sources are present in the environment, I found that a notable amount of energy at these frequencies were present at the locations chosen based on previous studies claiming that these frequencies may be present. With clear data to demonstrate public exposure to these frequencies, it now seems more imperative to work towards fully comprehending the physiological effects these frequencies have on human health and wellbeing, both positive and negative effects, so that this information could then be used in the planning and designing of urban spaces.

In: Movement 2019: Brain, Body and Cognition
Editors: Joav Merrick and Gerry Leisman
ISBN: 978-1-53616-714-6

Chapter 34

EMBODIED RHYTHMS: QUADRATO MOTOR TRAINING MAY BENEFIT CORTICAL FUNCTIONAL CONNECTIVITY IN PRODROMAL ALZHEIMER'S DISEASE: AN EEG STUDY

Claudio Babiloni, Claudio Del Percio Roberta Lizio, Francesco Orzi, Carla Buttinelli, Franco Giubilei, Antonio De Fano, Fabio Marson and Tal Dotan Ben-Soussan*

Department of Physiology and Pharmacology "Vittorio Erspamer",
University of Rome "La Sapienza", Rome, Italy;
Institute for Research and Medical Care, IRCCS San Raffaele Pisana, Rome, Italy;
Department of Neuroscience, Mental Health and Sensory Organs,
University of Rome "La Sapienza", Rome, Italy;
and Research Institute for Neuroscience, Education and Didactics,
Patrizio Paoletti Foundation for Communication and Development, Assisi, Italy

Keywords: sensorimotor training, movement, cognition, embodiment, Alzheimer's disease

Objective: Dementia is a growing epidemic disorder, which includes progressive neurodegenerative deterioration of memory, emotions, motor and social functions. Alzheimer's disease (AD) is the most common neurodegenerative cause of dementia. Since no disease-modifying drug exists, it is important to test non-pharmacological treatment options to slow down the disease progression. Previous studies have shown that

* Corresponding Author's Email: research@fondazionepatriziopaoletti.org.

exercise and dance therapy can have a significant positive symptomatic impact in improving cognitive, motor, and emotional functioning in people with dementia. In this vein, Quadrato Motor Training (QMT) is a specifically structured sensorimotor training that requires making one-step movements to different corners of the square in response to rhythmic verbal instruction. Importantly, QMT was found to enhance brain neuronal synchronization as revealed by electroencephalographic (EEG) rhythms in healthy adults, which is known to be altered in AD.

Method: Sixty amnestic mild-cognitive impairment (aMCI) subjects with possible prodromal AD were exposed to either 28-day QMT or a control walking motor training (WMT, similar to QMT but without the cognitive engagement). Standard neuropsychological tests, resting state eyes-closed EEG recordings, and motor assessments were performed before and after the training program. Functional cortical connectivity was tested by the popular eLORETA freeware for EEG source estimation.

Result: Compared to the WMT group, the QMT group showed enhanced functional connectivity from right occipital to right parietal and temporal source alpha (8-10 Hz) activity after the training program.

Conclusion: The present results indicate that QMT enhanced functional cortical connectivity in prodromal AD patients at 1-month follow up and motivate longer longitudinal experiments to test the hypothesis that it can slow down the neurodegenerative deterioration with beneficial effects at 3-12 month follow-ups. We are currently working in this direction.

In: Movement 2019: Brain, Body and Cognition
Editors: Joav Merrick and Gerry Leisman
ISBN: 978-1-53616-714-6

Chapter 35

QMT-INDUCED MOLECULAR AND COGNITIVE EFFECTS: THE CASE OF AUTISM

Fabio Marson*, Sabrina Venditti, Micaela Caserta, Loredana Verdone, Valerio Vetriani, Joseph Glicksohn and Tal Dotan Ben-Soussan

Research Institute for Neuroscience, Education and Didactics, Patrizio Paoletti Foundation for Development and Communication, Assisi, Italy;
Department of Biology and Biotechnology "Charles Darwin", Sapienza University of Rome, Italy;
Institute of Molecular Biology and Pathology, National Council of Research (CNR) Rome, Italy;
Department of Criminology, Bar-Ilan University, Raman Gan, Israel;
and The Leslie and Susan Gonda (Goldschmied) Multidisciplinary Brain Research Center, Bar-Ilan University, Raman Gan, Israel

Keywords: autism, QMT, cognition

Objective: Autism spectrum disorder (ASD) is a multi-system disorder that impacts on motor control, cerebellar coordination of movements and cognition. Among its core features are impaired social interaction, impaired language and motor dysfunctions. Among other theories, it has been suggested that these may be attributed to cerebellar dysfunction. The cerebellum is central in a wide range of functions sometimes found to be impaired in autism, including timing and coordination of movement, motor learning, evaluation of the match between intention and action, predictive learning, environmental exploration, behavioral inhibition, attention, and visual orienting. A possible rationale for cerebellar involvement in autism could lie in the cerebellum's role as an oscillator, producing synchronized activity within neuronal networks, including sensorimotor networks critical for cognitive, social and emotional skills. Healthy neuronal synchronization is related to cognitive and motor functions, as well as to creativity and emotional well-being. Quadrato Motor Training (QMT), a structured sensorimotor training that involves sequences of movements based on verbal commands, was found to increase cerebellar oscillatory alpha (10 Hz) activity and to improve reading in dyslexics. In addition, QMT was found to enhance creativity and to improve spatial and temporal cognition, which are known to be altered in ASD. QMT was further shown to induce neurotrophic changes, specifically concerning Brain Derived Neurotrophic Factor (BDNF), which is also known to be altered in autism. So far, no studies have examined the effects of QMT on autism.

Method: A longitudinal study examining the cognitive and molecular effects of QMT on a 9 year-old child suffering from ASD was conducted. Spatial and temporal cognition were measured using the Hidden Figure Test (HFT) and Time Production (TP) tests, respectively. Salivary proBDNF was also collected before and following 10 weeks of daily training.

Result: Following 10 weeks of training, the participant showed improved spatial and temporal cognition, as revealed by the HFT and TP tasks. We are currently analyzing the proBDNF.

Conclusion: The results confirm the link between specifically-structured sensorimotor activation and cognitive change, also in the case of ASD. QMT seems to be a useful tool for the treatment of different aspects of neurodevelopmental states, including dyslexia and autism.

In: Movement 2019: Brain, Body and Cognition
Editors: Joav Merrick and Gerry Leisman
ISBN: 978-1-53616-714-6

Chapter 36

DIAGONAL THINKING: EEG MODULATIONS OF DURING THE QUADRATO MOTOR TRAINING

Fabio Marson**, *Patrizio Paoletti and Tal Dotan Ben-Soussan
Research Institute for Neuroscience, Education and Didactics,
Patrizio Paoletti Foundation, Assisi, Italy

Keywords: EEG, sensorimotor training, QMT, diagonal movements, navigation

Objective: Quadrato motor training (QMT) is a whole-body sensorimotor paradigm involving both cognitive and motor processing. QMT is constructed by a sequence of one-step movements in one of three possible directions (i.e., vertical, horizontal and diagonal, VM, HM and DM respectively). In previous studies, QMT was found to increased cognitive flexibility, creativity and reading. MRI and EEG studies using a pre-post design demonstrated that increased structural and increased alpha connectivity could underlie these cognitive improvements. However, no study examined cerebral activity during the actual performance of QMT so far.

Method: In this preliminary study, nine volunteers performed QMT during which we recorded EEG signals. Volunteers trained on the QMT for 28 days prior to the EEG recording which was conducted using Eego sports system (ANT Neuro, Enschede, Netherlands). Data pre-processing and analysis were conducted on MatLab using EEGLAB toolbox. Event Related Potentials (ERPs) analysis has been performed comparing diagonal to vertical and horizontal movements.

Result: ERPs showed significant increased electrophysiological frontal activity during the planning (200 ms after instructions) and at the end of DM (700-800 ms).

* Corresponding Author's Email: research@fondazionepatriziopaoletti.org.

Moreover, we observed a similar increased activity over parietal regions during the execution of DM (500 ms).

Conclusion: Higher complexity of DM is indexed by the increased activity over frontal regions during movement planning and computation. Moreover, increased activity over parietal and frontal regions, respectively at 500 and 800 ms after instruction, represent feedback monitoring of movement execution and body position in space. Modulation of cerebral activity observed during computation, execution and monitoring of diagonal movements represent an important functional part of the QMT. Integration of these processes could promote connectivity in order to perform efficient computation of movements, linking together spatial, proprioceptive and cognitive processing.

In: Movement 2019: Brain, Body and Cognition
Editors: Joav Merrick and Gerry Leisman
ISBN: 978-1-53616-714-6

Chapter 37

QUADRATO MOTOR TRAINING EFFECTS ON CNV AND MOTOR PREPARATION: IMPLICATIONS FOR HEATHY AGING

Stefano Lasaponara, Antonio De Fano [*] ***and Tal Dotan Ben-Soussan***

Research Institute for Neuroscience, Education and Didactics, Patrizio Paoletti Foundation for Development and Communication, Assisi, Italy;
IRCCS Fondazione Santa Lucia, Department of Neuropsychology, Via Ardeatina, 306 Rome, Italy;
and Università LUMSA, Dipartimento di Scienze Umane, Rome, Italy

Keywords: aging, movement, sensorimotor training, QMT, CNV

Objective: Normal aging is typically associated with motor response slowing and motor performance impairment. EEG studies can closely examine the time course and frequency selectivity of brain activity related to various cognitive and motor-related processes, providing relevant indices of early functional changes associated with normal and pathological aging. The motor priming paradigm, in which a preparatory stimulus (S1) precedes the imperative stimulus (S2), has been widely employed to disentangle the phases of preparation and execution of movement. The main EEG parameter utilized to depict the time course of motor preparation processes during the S1-S2 interval is the contingent negative variation (CNV). Following the early wave related to S1 orienting activities, the late wave of the CNV is mainly composed of a fronto-central motor preparation component, with additional contribution of parietal nonmotoric stimulus expectation processes. Reports of age effects on the motor-related late CNV component has been described either reduced in old versus young subjects. Studies investigating the effects of training, such as meditation on the CNV in participants reduced CNV in choice reaction time task (CRT), and increased CNV in simple reaction time task (SRT), suggesting that meditation can induce a more balanced attentional state. In the current study, we tested whether a similar effect could be obtained in healthy non-meditators using a single session of a specifically-structured sensorimotor training (Quadrato Motor Training - QMT). QMT involves a combination of attention with motor response for producing the correct sequences of movement, and was found to improve coordination, attention and creativity.

Method: 23 healthy right-handed volunteers participated in the study. All participants performed the SRT and CRT tasks immediately before and soon after a single QMT trial

Result: We found that following a QMT session, CNV amplitude reduced in CRT and increased in SRT. Conversely, the P3 amplitude increased in CRT and decreased in SRT.

Conclusion: Taken together, these results support the idea that QMT has attentional benefits in normal healthy participants. QMT is a useful sensorimotor training across the lifespan. Future studies should be conducted in order to test the hypothesis that QMT can serve as a neuroprotective program.

In: Movement 2019: Brain, Body and Cognition
Editors: Joav Merrick and Gerry Leisman
ISBN: 978-1-53616-714-6

Chapter 38

OLD BRAIN-BODY DANCING

Daniela Aisenberg*
Ruppin Academic Center, Israel

Keywords: dance, old age, movement, brain

The relation between movement and aging will be presented via brain imaging studies performed in Italy. Then, a methodological model for dancing will be explained by a professional choreographer and dance teacher, followed by supporting research of cognitive and emotional effects. Last, a live dance-presentation will be performed by old-women dancers.

Normal aging is associated with motor response slowing and motor performance impairment. EEG studies allow examination of the time course and frequency selectivity of brain activity related to various cognitive and motor-related processes, providing relevant indices of early functional changes associated with normal and pathological aging. An important EEG parameter utilized to depict the time course of motor preparation processes is the contingent negative variation (CNV). Reports of age effects on the motor-related late CNV component has been shown reduced in old compared to young adults. Various methods for normalizing it are under examination. Dance is classified as one of the enriched environments known to induce an increase of neurotrophic factors, which promote brain plasticity. This may provide an explanation for motor improvements following dancing. In addition, physical activity can also affect the regulation of neurotransmitters. It is also known that body awareness is increased after participating in dance groups. Dance can enhance the concentration of serotonin, "the happiness hormone". In this way, dancing can make old adults feel better and enjoy their retirement.

* Corresponding Author's Email: danielaa@ruppin.ac.il.

In: Movement 2019: Brain, Body and Cognition
Editors: Joav Merrick and Gerry Leisman
ISBN: 978-1-53616-714-6

Chapter 39

PSYCHOLOGICAL AND COGNITIVE EFFECTS OF DANCING IN OLD AGE

Daniela Aisenberg*[*] *and Dana Kinory
Department of Clinical Psychology- Gerontology, Ruppin Academic Center, Israel

Keywords: dance, healthy aging, cognitive abilities, psychological well-being, longitudinal study

Objective: Dancing has been advocated as an effective intervention that improves cognitive, physical and emotional functions among elderly. Dance embodies many aspects, such as sensory and auditory stimulation, physical activity and cognitive challenges, rhythmic motor coordination, balance, musical experience, social interaction, memory, perception, and expression of emotions. Previous studies have shown that dance is a therapeutic tool for people who suffer from movement disorders, such as Parkinson's-disease, and depression. Among these patients, dance was found to improve balance, walking speed, gait and cognitive performance. Moreover, it was found beneficial for mood expressions, leading to improvement in life quality and well-being of the patients. Despite the growing potential and popularity of dance as therapy, there are limited numbers of studies to date that examine the benefits of dance for healthy old adults in the community over time. Therefore, the purpose of this study is to examine the emotional and cognitive long-term effects of dance among dancers aged 65-100 years.

Method: Old women from the "Gila" workshops of the choreographer Galit Liss took part in the study and are still under follow-up examination. Cognitive abilities were measured using the Simon task, a well-known response-inhibition task. Emotional

[*] Corresponding Author's Email: danielaa@ruppin.ac.il.

aspects were measured by self-report questionnaires; non-judge subscale from the Five-Facet-Mindfulness-Questionnaire (FFMQ), depression module of the Patient Health Questionnaire (PHQ-9). In addition, explicit report of ageism was tested using the Fraboni Scale of Ageism. All measures were screened at time 1, and again after several months, or following a highlight event (such as a dance show of the group). Demographic details were also collected, for further analysis.

Result: At time 1, dancers showed relatively high levels of cognitive performance, compared to previous studies of these age groups in this task. Specifically, high accuracy rates fast reaction times were found. Nevertheless, old age congruency effects were observed. Emotionally, screening at time 1 revealed mixed reports, also regarding ageism, partly related to demographic measures. Since the research is a longitudinal study, data are still collected and will be updated close to the conference.

Conclusion: Participating in dancing groups may positively affect cognitive and emotional aspects over time.

In: Movement 2019: Brain, Body and Cognition
Editors: Joav Merrick and Gerry Leisman
ISBN: 978-1-53616-714-6

Chapter 40

IMPLICATIONS FOR HEALTHY AGING

Stefano Lasaponara, Antonio De Fano [*]
and Tal Dotan Ben-Soussan
Research Institute for Neuroscience, Education and Didactics, Patrizio Paoletti Foundation for Development and Communication, Assisi, Italy;
IRCCS Fondazione Santa Lucia, Department of Neuropsychology, Via Ardeatina, Rome, Italy;
and Università LUMSA, Dipartimento di Scienze Umane, Rome, Italy

Keywords: aging, movement, sensorimotor training, QMT, CNV

Objective: Normal aging is typically associated with motor response slowing and motor performance impairment. EEG studies can closely examine the time course and frequency selectivity of brain activity related to various cognitive and motor-related processes, providing relevant indices of early functional changes associated with normal and pathological aging. The motor priming paradigm, in which a preparatory stimulus (S1) precedes the imperative stimulus (S2), has been widely employed to disentangle the phases of preparation and execution of movement. The main EEG parameter utilized to depict the time course of motor preparation processes during the S1-S2 interval is the contingent negative variation (CNV). Following the early wave related to S1 orienting activities, the late wave of the CNV is mainly composed of a fronto-central motor preparation component, with additional contribution of parietal nonmotoric stimulus expectation processes. Reports of age effects on the motor-related late CNV component has been described either reduced in old versus young subjects. Studies investigating the effects of training, such as meditation on the CNV in participants reduced CNV in choice

[*] Corresponding Author's Email: research@fondazionepatriziopaoletti.org.

reaction time task (CRT), and increased CNV in simple reaction time task (SRT), suggesting that meditation can induce a more balanced attentional state. In the current study, we tested whether a similar effect could be obtained in healthy non-meditators using a single session of a specifically-structured sensorimotor training (Quadrato Motor Training - QMT). QMT involves a combination of attention with motor response for producing the correct sequences of movement, and was found to improve coordination, attention and creativity.

Method: 23 healthy right-handed volunteers participated in the study. All participants performed the SRT and CRT tasks immediately before and soon after a single QMT trial

Result: We found that following a QMT session, CNV amplitude reduced in CRT and increased in SRT. Conversely, the P3 amplitude increased in CRT and decreased in SRT.

Conclusion: Taken together, these results support the idea that QMT has attentional benefits in normal healthy participants. QMT is a useful sensorimotor training across the lifespan. Future studies should be conducted in order to test the hypothesis that QMT can serve as a neuroprotective program.

In: Movement 2019: Brain, Body and Cognition
Editors: Joav Merrick and Gerry Leisman
ISBN: 978-1-53616-714-6

Chapter 41

THE YES MANIFESTO TO THE ELDERLY BODY 3: THE PHYSIOLOGICAL AESTHETIC OF THE ELDERLY NON-DANCER BODY

Galit Liss*
Independent choreographer, Tel Aviv, Israel

Keywords: dance, movement, elderly body, cognition, brain

Objective: The notion that the elderly body has its own aesthetic merits leads to exploring ways for constant growth and improvement in old age. This study raises the question of what is quality movement? How can non-skilled elderly dancers achieve quality movement? Here I'll provide a demonstrative presentation showing how do we build a movement language, syntax and meaning through this body. Through sensory, motoric and cognitive work, I hope to demonstrate the many ways in which an attentive brain-body mechanism can inspire personal growth, enable physical proficiency, and help an individual discover unique physical and artistic qualities.

Method: Over 450 non-dancer women (aged 60-83 years) participated in the "Gila" workshops dance groups. They were challenged with thinking patterns that inhibit the development of the body movement and its space perception. Using a methodology I developed, we created a dialogue of brain-body, enabling physical refinement and movement aesthetics.

Result: Creative processes were shown to bring different approaches to the body and patterns of thought in the context of the dancing body.

* Corresponding Author's Email: galit.liss@gmail.com.

Conclusion: Combining artistic language with principles of brain plasticity through sensory, motoric and cognitive work, creates a cognitive detour to physical limitations, and even leverage these disabilities into a singular physiological esthetics. An important consequence is challenging the subjective way of thinking of the elderly body.

Oral presentation

In: Movement 2019: Brain, Body and Cognition
Editors: Joav Merrick and Gerry Leisman
ISBN: 978-1-53616-714-6

Chapter 42

"GO" (SHORT PRESENTATION FROM THE SHOW)

Galit Liss[*]
Independent choreographer, Tel Aviv, Israel

Keywords: dance, adult body, performance, live show

Objective: A group of women, whose years are sketched upon their bodies, will perform a live show. While most of them danced as girls, as they matured, they were required, by their parents, to stop. These women returned to dance at an advanced age, challenging the common conception that sanctifies the younger body in general and the dancer's body in particular. "GO" is a manifesto for this body. The show premiered at the Suzanne Dellal Center in December 2017 and has received enthusiastic responses from audiences and critics.

Method: In this performance, moving side by side are a pilot, a ballerina, and an elderly woman. From the relationship which is built and unfolded between the images and the physical body, questions arise about presence, free will, coercion, and domination. "GO" seeks to expand the discussion beyond the particular body to our existence as individuals and as a society. The performance promotes the idea of living in a society that enables development and self-fulfillment in later life, even in an adult body.

Women dancers (aged 60-83 years), will perform a segment of 15-20 minute, from the performance "GO".

[*] Corresponding Author's Email: galit.liss@gmail.com.

In: Movement 2019: Brain, Body and Cognition
Editors: Joav Merrick and Gerry Leisman
ISBN: 978-1-53616-714-6

Chapter 43

AUTISM SPECTRUM: A NEW THERAPY INTERVENTION "LE REVEIL ACTIONS"

Piero Crispiani and Eleonora Palmieri*
Department of Special Pedagogy, University Macerata, Macerata,
Psychological and Pedagogical Victor Center, Macerata, Italy

Keywords: adaptation, executive functions, severe dyspraxia, neuro-activation

Objective: Autism spectrum under the lens of a spatial and temporal disorder in term of a severe dyspraxia, connected with a disharmonic functioning in different areas (motor, language, perception,).

Method: D.A.M.A (Dyspraxic Adaptive Motor Autism) is a professional program combining motor coordination and executive functions in an appropriate setting and a global approach. The therapy is based on some important actions that are defined " Le reveil actions" with the aim to improve important aspects of Autism Spectrum: adaptation timing, self-regulation and inhibitory processes. Motor and cognitive combination is the matrix of the intervention including many specific exercises to facilitate at the beginning a better neuro-activation in executive functions, improving the relationship and approach to children. Parents are involved in the training and they are co- therapist in the program.

Result *and Conclusion*: Practical examples will be presented and cases as a demonstration of intervention. Results will be reported during the oral presentation.

* Corresponding Author's Email: palmieri.eleonora1@gmail.com.

In: Movement 2019: Brain, Body and Cognition
Editors: Joav Merrick and Gerry Leisman
ISBN: 978-1-53616-714-6

Chapter 44

DECISION MAKING "ON THE MOVE": HOW CAN MOVEMENT INFORM OUR DECISIONS?

Milca Leon*
Department of Dance and Movement Therapy, School of Creative Arts Therapies, Haifa University, Haifa and Wingate College of Physical Education, Natanya, Israel

Keywords: cognition, decision making, intuition, movement, body

Modern life is imbued with countless decisions. Throughout life we are required to consider various factors on our way to make decisions regarding the more 'substantial' aspects of our life, e.g., personal life, family, work and career, and the more mundane aspects such as where to eat, what to buy as a present, and so on. Researches of decision making processes often focus on the influence and significance of the intuitive, cognitive and emotional aspects of the process, often overlooking the significance of the bodily and movement aspects, as if the 'decision-maker' was separate from his/her Body and movement at that time.

Objective: the objective of this workshop is to introduce, through personal experience, the role that the moving body has on our decisions, often without our conscious knowing; and to explore 'embodied decision-making process' and its value for the decision-maker. In this workshop we will experience decision making "on the move"' and will explore how the experience of our moving body can inform our decisions. We will explore how movement affects our 'knowing', and how we can learn to trust body and movement to inform our intuition and cognition of what is 'right' and worth considering at the time of deciding.

* Corresponding Author's Email: milcaleon@gmail.com.

The practical and theoretical background of this workshop is based mainly within the framework of Laban Bartenieff Movement Studies (LBMS), and the special area within the system that focuses on decision-making. Through the system we will experience and reveal our values as experienced in movement, our personal movement biases while making decisions, and some universal affinities between movement and decision making. This is an experiential workshop in which the theory will be revealed to the participants through the work.

In: Movement 2019: Brain, Body and Cognition
Editors: Joav Merrick and Gerry Leisman
ISBN: 978-1-53616-714-6

Chapter 45

TAKING TIME: ATTENTIONAL EXPERIENCES OF DANCERS WHO MOVE SLOWLY

Michelle Outram [*] ***and Clemens Wöllner***
ERC Project SloMo, Institute for Systematic Musicology,
University of Hamburg, Hamburg, Germany

Keywords: dance, slowness, attention, expertise, time

Objective: Slowness in dance is just beginning to emerge as an area of research in the cognitive sciences. Studies have so far focused on audience experiences. We aim to expand this research area to include performer experiences, beginning with our current study.

In sport and dance, movement sequences are learned consciously, but the automatisation of these sequences is almost always a part of the training process. Automatisation assists the performer in achieving a higher consistency, including at speeds faster than could be consciously achieved. We proposed that moving slowly may not involve these same processes of automatisation, but instead, attentional and physiological processes more similar to mindfulness.

We interviewed eight contemporary dance professionals with a specific practice of moving slowly to determine whether common experiences of slowness could be found and discussed in relation to research into attentional processes such as during meditation. Each practitioner has over 20 years of professional experience.

Result: Analysis of the interviews shows that the practitioners understand moving slowly in terms of taking time to pay more attention to sensory details and the choices

[*] Corresponding Author's Email: michelle.outram@uni-hamburg.de.

they are making, rather than as an aim in itself. This includes attentionally 'spotlighting' parts of their bodies, controlling the movement of their attention, and developing a greater awareness of interceptive processes such as heart rate and breathing. Practitioners also report connections to Eastern practices such as yoga, tai chi and zen.

Conclusion: Results suggest these practitioners may be training their attentional processes to the point where they are developing a type of attentional expertise. The existence of this type of expertise could challenge the accepted view that expertise is necessarily related to the automatisation of low-level rather than high-level processes such as attention allocation. In the context of the empirical study of dance, this would entail reassessing notions that sequence automatisation constitutes a defining element of dance. As the experiences described by the practitioners are somewhat alike to those of mindfulness meditation, research into slowness in dance through the frame of empirical studies into meditation may prove fruitful.

In: Movement 2019: Brain, Body and Cognition
Editors: Joav Merrick and Gerry Leisman
ISBN: 978-1-53616-714-6

Chapter 46

SENSE MAKING THROUGH MOVEMENT, DANCE AND COGNITION: THE MINDFUL CHOREOGRAPHER

Simone Kleinlooh*
Codarts University of the Arts and Erasmus University, Rotterdam, Netherlands

Keywords: dance movement therapy, choreography, cognition

In dance movement therapy improvisational dance and choreographic structures are ways to express emotions through body, movement and dance instead of words alone. Creating, shaping and forming movement material towards the creation of a meaningful dance-narrative requires the following set of actions; spontaneous moving, memorizing, deciding, observing, describing, analyzing, sensitizing, evaluating, and revising. This integrated approach addresses simultaneously the body, its movements and affects while cognitive processes as executive functioning support the forming and consolidation of a choreographic structure. It becomes a dance when what is perceived makes sense to the creator or the witnesses while generating aesthetic embodied responses. In the practice of dance movement therapy this approach has proven to be effective in many areas of mental health care and for different populations.

Improvisational dance and creation are fun, however can be perceived as uncomfortable when too many thoughts or feelings arise and consequently block the body and the creative flow. If this happens the art of mindfulness well known by dance movement therapists, is a supportive approach. Being mindful requires self-reflexivity, sensitivity and acceptance to one's bodily sensations and it takes us into the present

* Corresponding Author's Email: skleinlooh@codarts.nl.

moment which reinforces the regulation of too strong emotions and thoughts that may arise while we move, dance and create with others.

In this workshop participants without any dance experience are being guided through a firmly structured approach from the art of dance. Mindful mastery of the own body and its movements through interaction with other participants will evoke playfulness and creativity resulting in the creation of a meaningful choreography. Extra emphasis is put on the use of time, weight, space and imagery in the process of creation.

In: Movement 2019: Brain, Body and Cognition
Editors: Joav Merrick and Gerry Leisman
ISBN: 978-1-53616-714-6

Chapter 47

BACK AND FORTH: COMBINING MOVEMENT, COGNITION AND MOVEMENT NOTATION

Tali Ronen** *and Lilach Shalit
School for the Arts of Dance, Faculty of the Arts, Kibbutzim College of Education, Technology and the Arts, Tel-Aviv, Israel;
School of Physical Education and Sports Ohalo College, Katzrin;
and Department of Dance, Orot Israel Academic College of Education, Elkana, Israel

Keywords: coordination, theoretical movement learning, Eshkol-Wachman Movement Notation

Physical movement can contribute to academic learning and to the improvement of cognitive skills, in addition to its importance in motor development.

This poster focuses on two research studies, which examine the effect of physical exercise and other movement activities on different populations. Both studies used the movement and structural principles of Eshkol-Wachman Movement Notation (EWMN).

EWMN is a language based on an analytical method that defines a framework of essential units, required for the description of bodily movement in time and space.

The research conducted by Tali Ronen is a field study that examined an intervention program consisting of a combination of movement by means of EWMN and academic learning. The aim of the program was to improve the achievements of 3rd and 4th grade pupils in learning an academic subject, as they studied two topics in geometry: angles and symmetry. The study compared two experiment groups and two control groups. This was Mixed Methods Research which made use of three research tools. The findings of the study clearly shows that the achievements of the pupils who participated in the experimental groups were higher than those of the pupils who participated in the control groups.

The research study conducted by Lilach Shalit examined the effect of movement and coordination exercises, based on EWMN, on students who have Attention Deficit Hyperactive Disorder (ADHD). The study examined a movement intervention program of thirteen weeks duration in two different but related fields of knowledge: attention, a cognitive skill, and coordination, a physical skill. This research group was chosen because earlier studies have shown that physical exercise is likely to have a positive effect on people with ADHD. The findings of the study shows improved results according to the attention and coordination tests that were administered.

The two research studies show the power of movement education, in promoting learning.

In: Movement 2019: Brain, Body and Cognition
Editors: Joav Merrick and Gerry Leisman
ISBN: 978-1-53616-714-6

Chapter 48

THE ROLE OF IMAGERY, TOUCH, AND INTEROCEPTION IN ACTIVATING CORTICAL AND SUBCORTICAL CIRCUITS THROUGH SOMATIC MOVEMENT PRACTICES: A SCIENCE-BASED APPRAISAL

Amit Abraham*

Department of Medicine, Division of General Medicine and Geriatrics Emery University, Atlanta, Georgia and Department of Kinesiology, College of Education, University of Georgia, Athens, Georgia, United States of America

Keywords: cortex, somatic movement practices, brain, symposium

Somatic practices facilitate a wide range of cortical and subcortical brain circuits. This symposium brings together experts who will discuss four somatic-related topics—visual imagery, touch, movement, and interoception—while discussing their neuro-cognitive mechanisms and presenting innovative scientific ideas for capturing the multisensory nature of somatic practices.

Somatic, or consciously sensed, movement and cognition are inherently linked. As such, somatic movement is associated with neuroplasticity, including multiple somatosensory mechanisms. In this symposium, four experts within the field of somatic practices will present scientific literature and clinical experience regarding past and current research as well as new research paths. The goal of the discussions will be to

* Corresponding Author's Email: amit.abraham@emory.edu.

elaborate the multidimensional nature and effects of mindful movement on motor and non-motor aspects of human performance.

The field of somatic movement encompasses a range of educational and therapeutic systems and approaches (e.g., the Alexander Technique, the Feldenkrais Method, Body-Mind Centering, the Franklin Method, and Dynamic Embodiment.) Such approaches seek to cultivate conscious awareness in movement in order to improve and maintain human health and well-being. Accordingly, the underpinning mechanisms of somatic approaches include a dependence on numerous neuro-cognitive circuits, many of which are far from being fully understood.

With recent developments in fields such as embodied cognition and affective and relational neuroscience, it has become more possible to get a glimpse into, and even begin to correlate, the ways in which disparate conceptual frameworks of somatic movement systems share a common scientific framework. That said, it seems that although current scientific research conducted within the field of somatics indeed reveals parts of the mystery, significant areas and fields remain unknown. Therefore, fine-tuning research goals and paths and formulating clinical insights into research questions are inquiries that seem to be warranted steps towards pushing the field of somatics (movement) forward.

In: Movement 2019: Brain, Body and Cognition
Editors: Joav Merrick and Gerry Leisman
ISBN: 978-1-53616-714-6

Chapter 49

VISUAL MENTAL IMAGERY: BEYOND WHAT THE EYE CAN SEE

Amit Abraham*
Department of Medicine, Division of General Medicine and Geriatrics
Emery University, Atlanta, Georgia, United States of America and
Department of Kinesiology, College of Education,
University of Georgia, Athens, Georgia, United States of America

Keywords: mental imagery, visual imagery, cognition, mental, symposium

Mental imagery (MI) is the fundamental human cognitive skill of creating and using mental images and metaphors associated a wide variety of stimuli, such as the smell of a rose, the sound of the ocean, the sight of a sunset, or the sensation of floating in the air. Visual imagery (VI) is the subtype of MI that focuses on a visual aspect of an image, such as when athletes see themselves crossing the finish line first. Used widely with both sport and neurological populations for improving various aspects of performance, VI exhibits multiple similarities to actual visual perception. Specifically, overlapping brain activity between visual perception and VI has been shown to reach up to 90%, in that deficits in the former have been shown to result in corresponding deficits in the latter.

Accumulating scientific evidence points to the beneficial effects of MI on somatic-both motor and non-motor-performance in various populations: from elite dancers to people with Parkinson's disease. However, further understanding of the complexity and multi-faceted nature of VI could benefit both research and clinical-rehabilitative purposes.

* Corresponding Author's Email: amit.abraham@emory.edu.

Motor imagery and dynamic neuro-cognitive imagery (DNI; aka "The Franklin Method?) are two somatic MI approaches that rely on VI for enhancing self-awareness, retraining posture, and optimizing medical rehabilitation. In this talk, insights and views on scientific research and practical applications of VI will be presented. Further, recent findings regarding VI from studies conducted by the presenter on motor imagery and DNI in dancers and people with Parkinson's disease will be discussed.

In: Movement 2019: Brain, Body and Cognition
Editors: Joav Merrick and Gerry Leisman
ISBN: 978-1-53616-714-6

Chapter 50

NEURAL MECHANISMS ASSOCIATED WITH INTEROCEPTION AND MOVEMENT IN SOMATIC PRACTICES: PAST AND FUTURE STUDIES

Martha Eddy*
St Mary's College, Moraga, California, Montclair State University, Montclair, New Jersey and University of North Carolina, Greensboro, North Carolina, United States of America

Keywords: neuroscience, movement education, brain, symposium

In order to establish a strong context for understanding neuroscience research within the relatively new field of Somatic Movement Education and Therapy, the work of pioneers such as FM Alexander, Gerda Alexander, Irmgard Bartenieff (physiotherapist and movement psychotherapist), Moshe Feldenkrais PhD, Charlotte Selver, Ida Rolf PhD, and Milton Trager MD, is analyzed. This talk seeks to demonstrate the areas of the brain that have been impacted by these leading somatic practices as well as the types of research in somatic studies that are directed toward a deeper scientific understanding of the neural mechanisms of interoception and movement.

After 30 years of case study experience, an additional 5-7 years of systematic cross-case analysis determined the defining features of over 40 somatic movement systems. It was also revealed 1) that research studies use newly emergent methods inclusive of qualitative, quantitative, and mixed methodologies; 2) that reliability and validity measures are variable; and 3) that the quantitative research studies on neural stimulus and

* Corresponding Author's Email: drmarthaeddy@gmail.com.

response in somatic movement have been primarily focused on cortical stimulation of the visual-kinesthetic or verbal-kinesthetic experience. While the importance of the sub-cortical aspects of neuroplastic changes evident in case studies have been highlighted, there has been a serious limitation in substantiating research. For instance, declarations of socio-emotional change in behavior that are cited to be governed by the limbic system have rarely included neuroscience or neural imaging of limbic activity.

Furthermore, the brain structures involved in interoception include other even lower brain structures such as the reticular formation, basal ganglia, thalamus and geniculate bodies, midbrain, cerebellum, and brainstem, as well as the nerves themselves. The predominant thesis is that more research, as well as improved instrumentation and research methodologies, is needed for recognizing the specific roles of deep brain structures in human behavior—inclusive of kinetic, kinematic, and biomechanical relationships.

In: Movement 2019: Brain, Body and Cognition
Editors: Joav Merrick and Gerry Leisman
ISBN: 978-1-53616-714-6

Chapter 51

WHAT HAPPENS IN VAGAL DOESN'T STAY IN VAGAL: THE FELDENKRAIS METHOD'S® IMPACT ON PARASYMPATHETIC RESPONSE THROUGH TOUCH AND MOVEMENT

Elinor Silverstein*
Guild Certified Feldenkrais Practitioner, Private Practice,
Orange County, California, United States of America

Keywords: vagal, parasympathetic, touch, movement, symposium

This talk explores the ways in which the somatic movement practices of touch and movement from the Feldenkrais Method® activate neuroplasticity in support of self-regulation, and offers a design study that could assess these processes quantitatively. Practitioners of the Feldenkrais Method® guide clients' movement and self-awareness through touch to change the way the brain perceives itself. Clients also learn movement sequences to activate different neural pathways that calm and settle the nervous system.

In the presenter's clinical experience of hands-on contact and Awareness Through Movement® exercises, the vagus, or the 10th cranial nerve, is stimulated and heart rate variability (HRV) increases. Part of the parasympathetic nervous system, HRV, in turn, allows a person to recover from stress activation more efficiently. Additionally, the vagus nerve works in relationship to the amygdala, a center of processing emotional reactivity. In essence, then, the Feldenkrais Method® supports neural and biological mechanisms of resilience.

Feldenkrais practitioners work with people suffering from pain (in the back, knees, neck, and so on), due to misuse or sometimes neurological injuries. Many of these clients have not been helped by conventional medicine, but report improvements in pain levels, range of motion and balance through the Feldenkrais Method®.

Qualitative accounts carry limited value in a scientific context, so researching the neural and biological impact of touch and movement in quantitative terms is an important next step in building the research base for this work. A proposal for a research study on the impact of Feldenkrais Method® exercises on heart rate variability will be presented.

In: Movement 2019: Brain, Body and Cognition
Editors: Joav Merrick and Gerry Leisman
ISBN: 978-1-53616-714-6

Chapter 52

SOMATIC MOVEMENT, PHYSICAL EXERCISE, AND MEDITATION: EXPLORING SIMILAR AND DISTINCT BRAIN STRUCTURES AND COGNITIVE BENEFITS

Elisabeth Osgood-Campbell*
Harvard Graduate School of Education,
Cambridge, Massachusetts, United States of America

Keywords: movement, exercise, meditation, brain, cognition, symposium

By teaching mindful awareness of bodily states in motion and at rest, somatic movement practices weave together kinesthetic and mental activities. Substantial evidence indicates that motor and cognitive processes are functionally related and exert bidirectional influence on one another. Along these lines, numerous studies reveal that moderate physical activity and meditation practice correlate with improvements in executive function tasks. For example, both acute treadmill walking and mindfulness practice are associated with increased measures of working memory and attentional control.

Given that somatic movement blends physical activity with mindful awareness, this presentation investigates two possibilities: first, that conscious or somatic movement practices may activate both overlapping and separate brain regions as physical exercise and mindfulness practice, and second, that somatic movement practices may be associated with similar and also unique cognitive improvements.

* Corresponding Author's Email: elo277@mail.harvard.edu.

To begin to test these hypotheses, the author presents a tripartite research project: a cross-sectional study to examine how the thickness of the somatosensory cortex is correlated with the number of years of somatic movement practice (of the Feldenkrais Method or BodyMind Centering, for example), an MRI study to investigate changes in key brain structures associated with cardiovascular exercise and meditation practice (the hippocampus, amygdala, cerebellum and brain stem) after somatic movement activities, and an assessment of attentional control among preadolescent children after a somatic movement practice. Criteria for valid research methodology for such an undertaking (i.e., definition of a standardized dose of somatic movement activity, sample size, randomization, and active control group) will also be discussed.

In: Movement 2019: Brain, Body and Cognition
Editors: Joav Merrick and Gerry Leisman ISBN: 978-1-53616-714-6

Chapter 53

EARLY PARENTAL-ENVELOP EXPERIENCES SHAPE EMOTIONAL AND PHYSICAL POTENTIALS, ALLOWING TO ANTICIPATE AND SHAPE CURRENT AND FUTURE ABILITIES OF THE SELF TO COPE WITH CRISES

Michelle Specktor*

Keywords: dance, therapy, movement, motion, behavior

Objective: Understanding human individuality and explaining why identical twins, although sharing the same genes and environment, may cope with crises so differently.

Methods: By means of a unique twinship study in the area of dance movement therapy, and the movement observation system Epimotorics, a psychomotor model for the analysis and interpretation of emotive motor behavior, by Yona Shahar-Levy (www.epimotorics.com).

Results: The newly introduced twin type's identification, TT1 and TT2, embraces the potential for coping with crises among individuals in general, and between twin pairs in particular. This distinction offers a new approach of predicting how children would cope with crises, being integral experiences in the child's developmental stages. TT1 and TT2 characteristics evolve based on the newborn twin's emotional and functional development in the early years of their twinship lives.

This identification reveals possible parental guidelines for appropriate holding and encouraging of their twins in their early years according to the newly proposed twin type's identification, TT1 and TT2. This would enhance the twins' emotional capabilities and consequently would improve their coping potential with crises throughout life.

Conclusion: Early parental-envelop experiences shape the child's emotional and physical potentials, allowing to anticipate and shape current and future abilities of the self to cope with crises.

In: Movement 2019: Brain, Body and Cognition
Editors: Joav Merrick and Gerry Leisman
ISBN: 978-1-53616-714-6

Chapter 54

SHIRODHARA: AN AYURVEDA TREATMENT IN THE MANAGEMENT OF BRAIN BEHAVIOR DISORDERS

Gaurang Joshi and Bhavna Joshi*
Atharva Multi Specialty Ayurveda Hospital and Cancer Research Center,
Panchakarma and Skin Care Hospital, Rajkot, Gujarat, India

Keywords: Shirodhara, Ayurveda, psychosomatic disorders

Shirodhara is a form of Ayurvedic snehana procedure, wherein involves gently pouring liquids over the forehead. This procedure can be a poorvakarma (preoperative procedure) to main panchakarma or pradhan karma (main procedure) depending upon the condition of the patient. Shirodhara is mainly indicated for neurobehavioral and psychosomatic disorders. Recent investigations report that, Shirodhara is effective in management of condition like pediatric attention deficit hyperactivity disorder, cerebellar ataxia, essential hypertension, psychological symptoms of menopause, premature ejaculation, generalized anxiety disorder and insomnia. The exact mechanism of its mode of action in not known till date. But, preliminary reports suggest that, this procedure has a demonstrable anxiolytic effect evidenced by decrease in plasma noradrenaline and urinary serotonin excretion, significant decrease in rate of breathing, reduction in diastolic blood pressure and heart rate with lowered sympathetic tone.

In: Movement 2019: Brain, Body and Cognition
Editors: Joav Merrick and Gerry Leisman
ISBN: 978-1-53616-714-6

Chapter 55

CHANGE PROCESSES: NON-VERBAL SYNCHRONY AS A LYNCHPIN OF CLIENTS' ADAPTIVE EMOTION REGULATION IN COACHING

Tünde Erdös*[*]*, Erik de Haan and Joshua Wilt
Amsterdam Business Research Centre, Vrije Universiteit Amsterdam, The Netherlands;
Ashridge Business School, United Kingdom;
Case Western Reserve University, Cleveland, Ohio, United States of America;
Department of Organisational Development and Coaching, Vrije Universiteit Amsterdam, The Netherlands;
Ashridge Centre for Coaching, United Kingdom and
Department of Psychology, Case Western Reserve University, Cleveland, Ohio, United States of America

Keywords: non-verbal synchrony, whole-body movement, coaching presence research, adaptive emotion regulation, change process

Objective: Little is known about the non-verbal mechanisms by which change processes affect clients' self-regulatory competencies for effective coaching. The question when and why coaches should apply which coaching strategies toward strengthening clients' self-regulatory capacities and attitudes as prerequisites of effective change outcomes remains a black box in coaching. This is unfortunate, as it is essentially the clients who are the recipients of coaching and who engage in the changing. In particular, non-verbal synchrony (NVS) between coach and client has so far received no attention in coaching psychology. Our longitudinal quantitative study observed 150 coach-client pairs over a period of 6-8 months. It explored the extent to which NVS drives clients' self-regulatory capacities and self-directed goal-attainment in coaching.

Methods: We investigated the indirect effects of NVS and goal attainment using motion energy analysis (MEA), which has three unique features: a) measuring dynamic quality; b) measuring NVS objectively and automatically by a video-computer interface; c) including both simultaneous movement and time-lagged coordinated movement in a window of ± 5 seconds.

Results: The compound effect-size of coaching effectiveness correlated with NVS. Associations of affect-balance, self-efficacy and changes in experiencing and behaviour were in the range of medium effect sizes (r = .2 to .4) and pointed in the same direction: synchrony was positively associated with effectiveness. NVS correlated with less interpersonal problems (r = -.25 to -.35), and with a secure attachment style (r = .30 to .39).

Conclusion: Movement quantification (MEA) along with the statistical approach for NVS allow for an objective quantification of embodied processes in coaching, which are normally not consciously controlled by individuals. Using MEA as an objective quantification of NVS may help in further advancing research on emotion regulation in coaching psychology.

In: Movement 2019: Brain, Body and Cognition
Editors: Joav Merrick and Gerry Leisman
ISBN: 978-1-53616-714-6

Chapter 56

Effects of the Impact of Different Types of Training on Activities of Daily Living and Their Perception by Old Age People

Celia Cohen Barros* and Celia Pereira Caldas
Medical Sciences Graduate Program, Rio de Janeiro State University,
Rio de Janeiro, Brazil

Keywords: aged, aging, exercise, functional capacity, women

Objective: This study aims at evaluating and comparing the effectiveness of three types of training - aerobic, strength, and power - in improving the functional capacity of aged women as well as assessing their perception of the changes brought about by exercising.

Methods: This study is characterized by being interventional, carried out on a population of aged volunteers and a parallel control group, and performed in three stages: a preliminary evaluation, intervention and follow-up re-evaluation, and a qualitative research on the content analysis of the volunteer's perception of changes. 123 physically independent women of old age not engaged in physical exercises were studied. The volunteers were randomly stratified into four groups: Control Group (GC, n = 33) - who did no exercise; Aerobic Group (GA, n = 30) - who practiced low intensity resistance exercise combined with moderate intensity walking; Strength Group (GF, n = 30) - who trained high intensity resistance exercises; and Power Group (GP, n = 30) - who trained moderate intensity resistance exercises at high speed. All groups went through the following tests: Aerobic endurance - the 6-minute walk test (T6m); Maximum strength -

one repetition maximum (1RM); Power - Tendo Weightlifting Analyzer (TWA); and Agility - up and walk. GA-, GF-, and GP-volunteers underwent 24 sessions of training.

Results: With the exception of the GC-group, the other groups showed significant improvement in all tests. The comparison of the gains achieved by the GA-, GF-, and GP-groups in the T6m, power, and agility tests exhibited some differences, though not significant. The improvement reached by the GF-group in the 1RM test was considerably higher than that of the other groups. In the qualitative research, the volunteers' testimonies revealed the importance of training for the improvement of the physical, psychological, social, and cognitive aspects of their lives.

Conclusion: We conclude that the aerobic, strength, and power training carried out did improve the functional capacity of the old age volunteers; in the 1RM test, the GF-group stood out from the others. Coherence was observed between the quantitative results and the volunteers' perception regarding the changes produced by physical training.

In: Movement 2019: Brain, Body and Cognition
Editors: Joav Merrick and Gerry Leisman
ISBN: 978-1-53616-714-6

Chapter 57

EXPLORING THE NEURAL MECHANISM OF VOLITIONAL QUALITY: A BRAIN MORPHOMETRY STUDY ON NATIONAL-LEVEL ATHLETES

Gao-Xia Wei**, Ruoguang Si, Shu Zhang, Li-Zhen Chen, Ying Yao and Zhi Wang

Key Laboratory of Behavioral Science, Institute of Psychology, Chinese Academy of Sciences, Beijing, China; Sports Science Research Institute of the State Sports General Administration, Beijing, China; and Cardiff University Brain Research Imaging Centre, Cardiff University, Cardiff, Uinted Kingdom

Keywords: volition, brain structure, sports, athletes, fMRI

Objective: This study aimed at investigating the neural mechanism of volitional quality by comparing brain morphometry between national-level Chinese athletes and university students with Magnetic Resonance Imaging (MRI).

Methods: The participants were 38 healthy young adults, including 19 short track speed skating athletes from Chinese national team (age: 18.3 ± 1.5; 10 males, 9 females) and 19 university students as controls (age: 19.2 ± 1.2; 9 males, 10 females). The volitional quality was measured with Volitional Quality Questionnaire (BTL-LYZ1.0, Li, 2014). The brain morphometry-cortical thickness was measured with high-resolution T1 MR images. Preprocess of the structural images was done by FreeSurfer 6.0, which is implemented in the Connectome Computation System (CCS: http://lfcd.psych.ac.cn/ccs.html), a pipeline system for discovery science of brain. The parcel-wised cortical thickness was computed between groups based on Destrieux atlas by considering sex, gender, education and intracranial volume as covariates. Also, partial correlational analyses in athlete group were conducted for exploring the association between volitional quality and cortical thickness.

Results: Athletes had higher score in goal clarity ($p = .000$), continence ($p = .008$), belief confirmation ($p = .000$), persistence ($p = .000$), thoughts concentration ($p = .000$) and difficulty bearing ($p = .000$) than the controls. Athletes also showed greater cortical thickness than the controls in brain areas in the right hemisphere including cuneus, precuneus, inferior parietal lobule, rostral middle frontal gyrus, superior frontal gyrus and calcarine fissure (Desikan-Killiany-atlas-based parcel-wise analysis with a Bonferroni corrected $p = 0.05/68$). Additionally, significant correlations between cortical thickness with significant difference and scores of volitional qualities were detected, including the rostral middle frontal gyrus and thoughts concentration ($p = .004$), and inferior parietal lobule and self-fulfillment ($p = .049$).

Conclusion: It suggested that the athletes showed greater volitional quality. Greater cortical thickness in middle frontal gyrus and inferior parietal lobule might revealed neural substrates of volitional quality. It also indicated that sports training is a crucial path to develop individual's volitional quality by optimizing regional brain morphometries.

Oral presentation: Biography: Gao-Xia Wei, an associate professor at Institute of Psychology, Chinese Academy of Sciences. She is the principle of several grants from National Natural Science Foundation of China, Chinese Academy of Sciences, and Beijing Science and Technology Ministry, respectively. She is also honored corresponding reviewer of evaluation panel in National Natural Science Foundation of China as well as the evaluation expert in Beijing Natural Science Foundation. Her research is mainly focused on movement-related behavior and brain plasticity, which covered several topics: 1) mind-body practice and brain health; 2) neural mechanism underlying the effect of sports training on elite performance with brain imaging technology; and 3) intervention technology development and application in health and clinical practice. Her publication covers the areas of Tai Chi Chuan, abdominal breathing, acute aerobic exercise and sports training (short track speed skating, swimming, diving, etc.).

In: Movement 2019: Brain, Body and Cognition
Editors: Joav Merrick and Gerry Leisman
ISBN: 978-1-53616-714-6

Chapter 58

DIVERSE EXERCISES SIMILARLY REDUCE OLDER ADULTS' MOBILITY LIMITATIONS

József Tollár*, Ferenc Nagy, Mariann Moizs, Kinga Szita, Lianne MJ Sanders and Tibor Hortobágyi
Somogy County Kaposi Mór Teaching Hospital, Kaposvár,
Hungary and University of Groningen, University Medical Center Groningen,
The Netherlands

Keywords: aging, exercise, gait, balance

Background: Little is known about the comparative effectiveness of exercise programs, especially when delivered at a high intensity, in mobility-limited older adults. We compared the effects of 25 sessions of high-intensity agility exergaming (EXE) and stationary cycling (CYC) at the same cardiovascular load on measured and perceived mobility limitations, balance, and health-related quality of life (QoL) in mobility-limited older adults.

Methods: Randomized to EXE (n = 28) and CYC (n = 27), mobility-impaired older adults (age 70) exercised 5x/week for 5 weeks at 80% of age-predicted maximal heart rate. Wait-listed controls did not exercise (n = 28).

Results: Groups did not differ at baseline in any outcomes ($p > 0.05$). The primary outcomes (SF-36: EXE: 6.9%, effect size: 2.2; CYC: 5.5%, 1.94; WOMAC: EXE:-27.2%, -3.83; CYC:-17.2, -2.90) improved similarly ($p>0.05$). Secondary outcomes, including body mass (-3.7%), depression (-18%), and walking capacity (13.5%) also improved ($p < 0.05$) similarly after the two interventions. Activities of daily living, Berg Balance Score, BestTest scores, and dynamic gait index improved more ($p < 0.05$) after EXE than CYC. Center of pressure of standing sway path improved in 1 of 6 tests only after EXE ($p < 0.05$). Post-exercise cardiovascular response improved slightly in EXE. CON did not change in any outcomes ($p>0.05$).

Conclusion: When matched for cardiovascular and perceived effort, vastly different high intensity exercise improved health-related QoL, perceived mobility limitation, and walking capacity similarly and balance outcomes more in mobility-limited older adults, expanding these older adults' evidence-based exercise options to reduce mobility limitations.

In: Movement 2019: Brain, Body and Cognition
Editors: Joav Merrick and Gerry Leisman
ISBN: 978-1-53616-714-6

Chapter 59

VASTLY DIFFERENT EXERCISE PROGRAMS SIMILARLY IMPROVE PARKINSONIAN SYMPTOMS: A RANDOMIZED CLINICAL TRIAL

József Tollár**, Ferenc Nagy, Mariann Moizs, Kinga Szita and Tibor Hortobágyi
Somogy County Kaposi Mór Teaching Hospital, Kaposvár,
Hungary and University of Groningen, University Medical Center Groningen,
The Netherlands

Keywords: exercise specificity, balance training, posture, quality of life

Objectives: To directly compare the effects of agility exergaming and stationary cycling exercise training on Parkinson's disease (PD) patients' mobility and clinical symptoms.

Methods: Design Randomized clinical trial. Setting Outpatient physiotherapy clinic in a hospital.

Participants Seventy-four, stage 2-3, none-demented PD patients. Intervention Exergaming-agility (EXE, n = 25), seated cycling (CYC, n = 25), and a wait-listed control group (CON, n = 24). EXE and CYC exercised 5x/week for 5 weeks matched at 80% of age-predicted maximal heart rate. Main outcomes for primary outcome: Movement Disorders Society-Unified Parkinson's Disease Rating Scale (UPDRS-II). Secondary outcomes: Parkinson's Disease Quastionnaire-39 (PDQ-39), Beck Depression Inventory (BDI), Schwab and England Activities of Daily Living (SE-ADL), Euro-Quality of Life-5 Dimension (EQ-5D), Berg balance scale (BBS), Balance Evaluation

Systems Test (BESTest), Tinetti Assessment Tool (TAT), Dynamic gait index, six-minute walk test (6MWT) and standing post urography.

Results: After treatment, UPDRS-II improved (mean change for EXE: -4.5, CYC: -3.2 points), PDQ (13%, 17%), BDI (-2.5, -2.1 points), 6MWT (129.6, 141.6 m), and EQ-5D (12, 9%) (all $p < 0.05$ but not different between groups). EXE vs. CYC improved SE-ADL (8.4, 4.0 points, effect size [ES]: 0.12), BBS (8.8, 4.2 points, 0.44), and 2 measures of post urography (ES: 0.11, 0.21) more ($p < 0.05$). BESTtest, TAT, Dynamic Gait Index, and 4 of 6 post urography measures did not change ($p > 0.05$).

Conclusion: Two highly different exercise programs improved most of the motor and clinical symptoms similarly in PD patients.

In: Movement 2019: Brain, Body and Cognition
Editors: Joav Merrick and Gerry Leisman
ISBN: 978-1-53616-714-6

Chapter 60

EXECUTIVE FUNCTION PERFORMANCE AND PREFRONTAL ACTIVATION WHEN CYCLING ON AN ACTIVE WORKSTATION IN YOUNG ADULTS

Tao Huang*[*], *Qian Gu, Zhangyan Deng, Zuosong Chen and Kun Wang
Department of Physical Education and Graduate School of Education,
Shanghai Jiao Tong University, Shanghai, China

Keywords: executive function, physical activity, active workstation, fNIRS

Objective: Previous studies suggested that exercise on an active workstation while working may be a promising intervention strategy for reducing sedentary time at workplace. The influence of using active workstation on cognitive function is not well studied and the existing studies yielded mixed findings. The purpose of the study was to examine the effects of self-paced cycling on an active workstation on executive function and prefrontal cortex activation in young adults.

Methods: In a cross-over study design, 35 young adults (mean age = 21.4 ± 2.6 years, 45.7% females) were randomly assigned to the following two task conditions separated by 48 hours: performing cognitive tests while sitting (SIT) and performing cognitive tests while cycling on an active workstation (ACTIVE). Executive function was assessed by a task-switching paradigm and Stroop Color and Word Test (SCWT) programed using E-Prime 2 professional (Psychology Software Tools, Inc., Sharpsburg, PA, USA), respectively. Stroop effects, global switch costs, and local switch costs were derived and used as the behavioral outcomes of the two tasks. Prefrontal activation was monitored using a 38-channel fNIRS system (NIRx Medical Technologies LLC, USA).

Results: The behavioral results showed that there were no significant differences on Stroop effects (112.61 ± 136.66 vs. 136.57 ± 122.24, $P = 0.19$) between SIT and ACTIVE conditions. The global switch costs (463.19 ± 35.47 vs. 452.77 ± 28.65, $p = 0.73$) and local switch costs (-6.14 ± 29.88 vs. 9.97 ± 26.77, $P = 0.70$) also did not differ. For the fNIRS results, compared with SIT, the oxy-Hb in response to Stroop interference in Channel 5, 10 and 12 was decreased during the ACTIVE condition (Ch5:0.40 ± 1.25 vs. -0.21 ± 0.88, Ch10: 0.41 ± 0.88 vs. -0.05 ± 0.72, Ch12: 0.31 ± 0.77 vs. -0.11 ± 0.58, all $P < 0.05$). The three channel was located at frontopolar area and left-dorsolateral prefrontal cortex (left-DLPFC). The oxy-Hb in all channels related to task-switching was not different between the conditions ($P>0.05$).

Conclusion: The results suggest that the behavioral performances on Stroop task and task-switching were not influenced by cycling on an active workstation; however, the cognitive resources for performing executive function may be reduced.

In: Movement 2019: Brain, Body and Cognition
Editors: Joav Merrick and Gerry Leisman
ISBN: 978-1-53616-714-6

Chapter 61

CORTICAL ACTIVATION DURING WALKING WHILE SMARTPHONE TEXTING: A DUAL TASK BASED FNRIS STUDY

Kun Wang**, Zhangyan Deng, Jimeng Zhang, Qian Gu, Tao Huang and Zuosong Chen
Department of Physical Education, Shanghai Jiao Tong University, Shanghai, China

Keywords: physical activity, walking, smartphone texting, dual task, fNIRS

Objective: The purpose of the study was to investigate the cortical activation during a dual task of walking while smartphone texting in young healthy adults using the functional near-infrared spectroscopy (fNIRS).

Methods: In a crossover study design, 39 right-handed college students (21.3 ± 2.5 years, 46.1% females) randomly undertook the following three task conditions separated by a minimum of 48 hours: smartphone texting only (T task), walking only (W task), and dual task of smartphone texting while walking (TW task). Cortical oxygenation during the three tasks was monitored using a 38-channel fNIRS (NIRx Medical Technologies LLC, USA). Walking was conducted on a treadmill with a speed of 2.0 km/hour. Texting task was performed with a typing APP on a smartphone.

Results: There was no significant difference in texting speed between T task and TW task (71.7 ± 10.4 vs. 71.0 ± 12.9 chars/min, $P > 0.05$). Texting enhanced hemodynamic response in frontopolar area (e.g., Ch1T task vs. W task, 1.10±1.93 vs. -0.12±0.14 a.u, $P < 0.01$; Ch1TW task vs. W task 0.89±1.05 vs. -0.12±0.14 a.u, $P < 0.01$) dorsolateral prefrontal cortex (e.g., Ch6T task vs. W task,0.24±0.43 vs. -0.15±0.42 a.u, $P < 0.01$; Ch6TW task vs. W task 0.21 ± 0.33 vs. -0.15 ± 0.42 a.u, $P < 0.01$) and Broca’s area (e.g.,

Ch35TW task vs. W task 0.99 ± 0.81 vs. 0.47 ± 0.75 a.u, $P < 0.01$). In addition, W task evoked an increased activation in temporopolar area (e.g., Ch8 T task vs. W task 0.04 ± 0.12 vs. 0.75 ± 1.30 a.u, $P < 0.01$) and superior temporal gyrus (e.g., Ch15 T task vs. W task 0.22 ± 0.64 vs. 0.76 ± 0.84 a.u, $P < 0.01$) than T task. There were no significant differences in those regions between T task and TW task.

Conclusion: The findings indicated that walking on a low speed requires less cognitive resources from the prefrontal cortex, while the temporal lobe is involved. When walking while texting on a smartphone, the brain areas (temporopolar area and superior temporal gyrus) involved in gait were activated, and areas in prefrontal cortex were also activated. Thus, more cognitive resources were allocated to smartphone texting during the dual task.

In: Movement 2019: Brain, Body and Cognition
Editors: Joav Merrick and Gerry Leisman
ISBN: 978-1-53616-714-6

Chapter 62

VIRTUAL REALITY TRAINING TO IMPROVE UPPER LIMB MOTOR FUNCTION IN MULTIPLE SCLEROSIS: A FEASIBILITY STUDY PROTOCOL

Alon Kalron**, *Michael Levy, Lior Frid, Hani Baransi, Mark Dolev, Maria Bovim-Rabey, David Magalashvili and Anat Achirona
Multiple Sclerosis Center, Sheba Medical Center, Tel Hashomer,
Department of Physical Therapy, Sackler Faculty of Medicine, Sagol School of Neurociences, Tel-Aviv University, Tel Aviv an, cVRHealth Ltd, Tel Aviv, Israel

Keywords: virtual reality, multiple sclerosis, cognitive

Objective: Evaluate the safety and feasibility of VR motor and cognitive training with the innovative Oculus-rift device in PwMS.

Methods: A prospective feasibility study was conducted including 30 PwMS with weakness in upper extremity. Patients practiced two VR games that require arm and shoulder movements in various directions, including reaching forward toward moving targets. Following the VR training session each participant completed a questionnaire covering aspects of quality of the training and fatigability. Additionally, the physiotherapist in charge of the training session completed a report assessing exercise performance, fatigability and safety of the training. Feasibility to perform the VR accurately and any intervention-related adverse events recorded.

* Corresponding Author's Email: omer@vrhealthgroup.com.

Results: Patients and examiner reported overall significant satisfaction regarding the VR training session and experience. Minimal adverse effects were recorded. Examiner reported that the VR training provided a suitable solution for PwMS.

Conclusion: VR is feasible and safe to use for PwMS. Assessment of VR training in PwMS in this feasibility study can help to design further targeted VR practicing to improve upper limb dysfunction in MS. Stage II will focus on implementing this intervention approach into a home-based environment.

In: Movement 2019: Brain, Body and Cognition
Editors: Joav Merrick and Gerry Leisman
ISBN: 978-1-53616-714-6

Chapter 63

FROM TOP-DOWN TO BOTTOM-UP OR THE OTHER WAY AROUND

Karni Ishai*
Private Practice, Mevaseret Zion, Israel

Keywords: body, soul, movement, Jung

The basic assumption of this workshop is that the body is the mirror of the soul. In many respects, the body and the soul are but one undividable holistic system containing complex physical and psychic processes which occur simultaneously and nurture each other. Movement therapy as a profession is based on the above assumption and provides a basis for various therapeutic models for a wide range of mental, cognitive, emotional and behavioral problems as well as interpersonal ones. It also offers tools for growth and development. One of the models combines this holistic body and soul approach with 'Active Imagination', a notion developed by Carl G Jung.

In their early stages, both movement therapy and the "Active Imagination" method had based themselves on the knowledge and healing forces of ancient cultures. Foundation to this knowledge is being provided nowadays, due to updated brain researches, and indeed this workshop is based on various brain research findings such as:

The "Plastic Brain" discovery, according to which the plasticity of the brain enables it to continue to change and acquire new skills throughout life by means of top down; "mirror neurons" that lays the neurophysiological foundations for empathy, compassion, interpersonal relationships and human relations in their deepest sense; the "social nervous system" and the concept of 'connectedness'.

In this workshop, I will invite the participants to experience three main aspects of movement therapy combining "Active Imagination":

The power of beneficial interpersonal relations which enable inner observation (introspection) and self-discovery.

The power of mirroring through symbolic movement in expanding the awareness.

The power of multi-sensory imagery translated into movement which combines Top-Down and Bottom-Up.

This workshop will interweave movement experience in the personal and interpersonal realms, together with additional channels that bypass consciousness. No movement background is required but rather an openness and a curiosity for movement experience all within the safe frame of the group.

In: Movement 2019: Brain, Body and Cognition
Editors: Joav Merrick and Gerry Leisman
ISBN: 978-1-53616-714-6

Chapter 64

THINKING ON YOUR FEET: DECISION MAKING IN DANCE CREATING

Milca Leon*

Department of Dance and Movement Therapy, School of Creative Arts Therapies, Haifa University and Wingate College of Physical Education, Natanya, Israel

Keywords: dance, cognition, decision making, creativity, intuition

Art creating, and more so dance creating, is a process of making decisions. A contemporary dance choreographer makes endless decisions regarding the movement composition, the dancers, music, costumes, set designs, lighting, etc. Like many people who strive in the "fast and forward' culture of current times, the choreographer-entrepreneur-producer, is required not only to make ample choices and decisions, but also to make them "on the move", i.e., immediately and within an ever changing dynamic environment. Being able to make artistic (and life) decisions "on the move" requires one to integrate perception, feelings, intuition and cognition in a complex 'dance' that takes place "here and now". It involves what the body and heart 'feel' is the 'right' decision with what the mind thinks is the 'correct' one; intuition and cognition. An important question arises then, are these capacities of cognition/intuition separate entities? Are they connected, and if so – how, and in what ways can this connection be developed? In my experience, within the dance-making world, the Body/Mind dichotomy is still 'alive', often favoring the Body (sensations, feelings, intuition) over the Mind (cognition, analysis). The famous dance critic Roger Copeland named it: "the revenge of the body". So how do we treat this rift?

Objective: The objective of this workshop is to introduce and experience a 'mini-dance creating' process and explore ways of revealing our "informed intuition", as a basis for making artistic decisions. Using both theory and practice, we will connect to our ability to "think-feel on the move" and explore our 'intuitive knowing', as a support for the myriad decisions of art (and life). The practical and theoretical aspects of this workshop are based within the framework of Laban Bartenieff Movement Studies (LBMS) which is a body of Body-Mind knowledge that is substantially applied within the field of Dance and Dance Education, and is significant in developing both artistic and analytic abilities. The practice will be supported by examples of using such processes with young Israeli dancers within the Dance Education system in Israeli high schools as a means of nurturing both their artistic and personal growth.

In: Movement 2019: Brain, Body and Cognition
Editors: Joav Merrick and Gerry Leisman
ISBN: 978-1-53616-714-6

Chapter 65

BRIDGING THE FEELING/THINKING DICHOTOMY IN CHOREOGRAPHY TEACHING

Milca Leon[*]
Department of Dance and Movement Therapy, School of Creative Arts Therapies,
Haifa University and Wingate College of Physical Education, Natanya, Israel

Keywords: dance, cognition, creativity, intuition

The inseparable nature of the human body and the human dance makes the art of dance a form of art which is 'closest' to the human bodily-experience. Writers, musician, painters, actors, directors - all have a medium, separate of them-selves, through which they create. Yet, dance, in most cases, is still being created 'here and now', often literally on its maker's own body. That makes dance more susceptible to the Feeling/Thinking dichotomy, usually being considered as 'subjective', 'intuitive', a 'spontaneous eruption of feelings and emotions expressed through movement'. Therefore, dance creating tends to be considered as a subjective process which is difficult to discuss or analyze, let alone teach.

As a choreography teacher, I often encounter, from both students and peers, the subjective approach, claiming that the intention of the choreographer is personal, therefore cannot be understood by the others, who are invited to "get their own meaning". While this is very democratic and noble of the dance people, it is also a way out from being accountable to one's own artistic intentions. I believe this does bad service to dance as a form of art, and it definitely makes the teaching of choreography hard.

In this lecture I would like to present an informal qualitative research taken in my work as a choreography teacher in Arts high-school in Haifa, and discuss ways of bridging over the Feeling/Thinking Dichotomy still prevalent within the dance education world in Israel.

In: Movement 2019: Brain, Body and Cognition
Editors: Joav Merrick and Gerry Leisman
ISBN: 978-1-53616-714-6

Chapter 66

SEEING THE RHYTHM WITHIN THE MOVEMENTS: HOW ATTENTION MODULATES THE OBSERVATION OF WHOLE-BODY MOVEMENT SEQUENCES

Shiau-Chuen Chiou * ***and Thomas Schack***
Neurocognition and Action Research Group, Center of Excellence Cognitive Interaction Technology (CITEC), Bielefeld University and Faculty of Psychology and Sports Science, Bielefeld University, Bielefeld, Germany

Keywords: action observation, attention, spatial and temporal processing, whole-body movement

Objective: Temporal information of movements, such as rhythm, is literally conveyed through spatial information, i.e., trajectory. However, whether the spatial information per se is also encoded when only the temporal information is in focus remains unclear. In this study, we aimed for understanding how the focus of attention on different aspects of information influences the visual processing of whole-body movement sequences.

Methods: Participants (n = 38) performed a delayed discrimination task (same/different judgment) with two different foci of attention: 1) Temporal-only, in which they attended to the temporal information of movements, as the spatial information would be kept constant. 2) Both, in which they attended to both the temporal and the spatial information to detect a change in either domain. Besides, a surprising movement recognition task (old/new judgment) was conducted at the end to test whether the spatial information was retained in the memory.

Results: Participants' sensitivity to the temporal change did not show significant difference between Temporal-only and both conditions ($p = .581$), suggesting that additional spatial processing did not impair temporal processing. Nevertheless, movement units were recognized better in the surprising recognition task when they were displayed in both than in Temporal-only condition ($p = .006$), illustrating a perceptual modulation from attention. A higher-than-chance recognition also indicated that the spatial information was encoded even when only the temporal information was in focus. Besides, there was a learning effect ($p = .018$) when participants started with Temporal-only condition, followed by Both condition (but not vice versa), showing that a better sensitivity to temporal features could be developed through an appropriate arrangement of training.

Conclusion: As participants' performance in the spatial discrimination task was pretty high, the additional cognitive load might be too low to induce an observable effect on the temporal processing, and thus the current results might be further modulated by the task difficulty. Altogether, spatial information of movements is processed during action observation, albeit not intentionally, and modulated by the focus of attention. Through a proper design, the sensitivity to temporal features can be developed, shedding light on practical applications in motor learning and therapy.

In: Movement 2019: Brain, Body and Cognition
Editors: Joav Merrick and Gerry Leisman
ISBN: 978-1-53616-714-6

Chapter 67

Conceptual Presentation of Dance Pedagogy Conflicts within the Dance Realm: Dance as a Meeting Point between Two Cultures?

Susan R Koff*
Dance Education at NYU/Steinhardt, New York, United States of America

Keywords: dance, movement, culture, expression

There are those who say that dance, as a non-verbal form of expression, is a perfect meeting place for those of other backgrounds and languages. It is a romantic notion. In reality, those systems that perpetuate western versions of dance tend to create a hierarchy of western forms over forms of non-western lineages. It is evident within formal education institutions of dance when those with a non-western form of dance are told they are deficient in "technique" that they need to be successful. Technique becomes code for western forms of dance in which balletic lines and capabilities are emphasized. In considering hierarchies, some challenges the western domination of music within music education, as I challenge the western domination of dance within dance education.

When pedagogy in dance is a meeting point between cultures, we need to recognize that dance knowledge, skills, worldviews, and civilizations from nonwestern cultures are not the same as those from western cultures. Classes need to be reconsidered in order to be open spaces for pedagogies of all cultures, in order to create a true meeting point open to sharing and exchange.

Bruner's conception that pedagogy communicates emotional, physical, social, cognitive, and rational learning processes of and to the learner and carries its own meanings, is what we need to recognize in order for dance pedagogy to be the true meeting point between two cultures.

In: Movement 2019: Brain, Body and Cognition
Editors: Joav Merrick and Gerry Leisman
ISBN: 978-1-53616-714-6

Chapter 68

MOTORIC ABILITIES ASSESSMENT OF PRESCHOOL AGE CHILDREN WITH INTELLECTUAL DISABILITIES

Snezana Nikolic**, Danijela Ilic-Stosovic and Nikola Babic
Faculty of Special Education and Rehabilitation, University of Belgrade and Higher Education School of Professional Health Studies in Belgrade, Belgrade, Serbia

Keywords: intellectual disability, preschool age, motorical performance

Objective: The aims of this paper were determining of discrepancy extensiveness in motoric functioning among children with intellectual disabilities and their peers of preschool age. After that, next objectives were determining the area of motoric functioning in which the discrepancy is most pronounced, and identifying the differences in motoric functioning regarding sex and age.

Method: The testing of the subjects using Educational test of motor functioning, a review of literature dealing with the subject of motoric functioning and knowledge systematization were the base on which this research was conducted on. SPSS descriptive statistics was used for the data analysis.

Results: The analysis of test results indicate the presence of the largest discrepancy between children with intellectual disabilities and their peers with typical development in the area of balance, then gross motor skills, followed by coordination and eye-hand coordination, and in the end, in the area of fine motor skills. Regarding the age of the children, by using Pearson's correlations, deviations in the performing of certain tasks in the field of balance (One foot standing: $r = -0{,}52$, $n = 30$, $p < 0{,}01$; Ankle pronation in one foot standing: $r = -0{,}37$, $n = 30$, $p < 0{,}05$), and in the field of gross motor skills (Catching and tapping a ball: $r = -0{,}44$, $n = 30$, $p < 0{,}05$; Catching the ball carefully: $r = -0{,}37$, $n = 30$, $p < 0{,}05$) were found. Regarding the gender, differences exist in the specific tasks in the area of balance (Jumping: $r = -0{,}38$, $n = 30$, $p < 0{,}05$) and coordination (Climbing: $r = -0{,}43$, $n = 30$, $p < 0{,}05$), in which boys demonstrate better developed skills.

Conclusion: Balance and Gross motor skills are shown to be the areas of motorical functioning with a most pronounced discrepancy between intellectually disabled children and their preschool age peers.

In: Movement 2019: Brain, Body and Cognition
Editors: Joav Merrick and Gerry Leisman
ISBN: 978-1-53616-714-6

Chapter 69

USING DEVELOPMENTAL PROFILE 3 SCREENING TEST TO ASSESS EFFECTIVENESS OF SENSORY-MOTOR-PERCEPTUAL PROGRAM FOR 5 AND 6 YEARS OLD CHILDREN

*Maia Vassileva**
KindyROO Europe

Keywords: sensory-motor-perceptual program, standardised tests, child general development, education

Objective: The objective of the research was to measure the impact of sensory-motor-perceptual program in comparison with standard physical education (PE) lessons and the effect of both programs on general child's development. We also wanted to evaluate is Developmental profile 3 (DP3) a test battery that can support data collection and analysis.

Method: We have obtained parental consent and worked with 241 five and six year olds in 6 pre-schools in Sofia, Bulgaria. The project lasted one academic year and was self-funded. Standardised for Bulgaria version of Developmental profile 3 was used as pre and post testing method. The children were split randomly in controlled and experimental groups and each group were assigned either a three day a week or five day a week training. The controlled groups followed the physical education curriculum of the Bulgarian Ministry of Education and the experimental groups followed "Unlock the Potential" (UP), an Australian, structured, sensory-motor-perceptual program. Parents filled in the DP3 questionnaires. The pre and post test results were processed with a statistical computer program SPSS.

Results: Data shows that both age groups, irrespective of intensity of training achieved higher scores in all five skill scales and in general development. The controlled group that did 3 and 5 days a week achieved the expected development level for the period. Statistically significant is only the increase in the experimental groups, where UP program was taught. The five days a week training achieved the highest improvements.

Conclusion: UP program has significant educational benefits. DP3 is a testing battery that can be used as the assessments for each of the five scales of child development: communication, social-emotional, adaptive behaviour, emotional, cognitive, physical gives opportunity to conduct detailed analysis. The test is standardised in number of countries and can be used for an international study thus assessing the national and cultural specifics that can or may affect childhood development.

In: Movement 2019: Brain, Body and Cognition
Editors: Joav Merrick and Gerry Leisman
ISBN: 978-1-53616-714-6

Chapter 70

CHANGING HABITUAL MOVEMENT AND THOUGHT PATTERNS USING ALEXANDER TECHNIQUE AND DANCE: A LECTURE PERFORMANCE

Sigal Bergman*
The Jerusalem Academy for Music and Dance, Jerusalem, Israel

Keywords: Alexander technique, changing habits, body awareness, injury rehabilitation

Objective: To show how changing thought can shift habitual movement and thought patterns. Use personal history to demonstrate how the process of becoming aware of habit, stopping it and replacing it with a new thought can lead to recovery from injury.

Methods: Lecture with physical demonstrations of content using dance and gesture. Exposure of automatic physical patterns by deconstruction. Guided imagery session.

Results: Audience becomes aware of the way they sit, and what they do with their head and neck as they observe the lecture demonstration. Discovering the possibility of changing physical habits using thought and imagery; specifically how when the head leads the spine movement is more efficient, and how when we let our thought connect our body to the space around us we undo compression in the body.

Conclusion: We are able to change habits of a lifetime in a moment if we put our mind to it. But we need to do it again and again in order to create new neurobehavioral patterns. This can help avoid injuries and overcome the ones we have.

In: Movement 2019: Brain, Body and Cognition
Editors: Joav Merrick and Gerry Leisman ISBN: 978-1-53616-714-6

Chapter 71

BODY-MIND DANCING AS AN INTERSECTION FOR TWO CONFLICTS: TECHNIQUE VS IMPROVISATION AND BEGINNING VS ADVANCED SKILL LEVELS

Martha Eddy*
BodyMind Dancing of Dynamic Embodiment Somatic Movement Therapy,
New York, United States of America

Keywords: embodied conflict resolution, somatic dance, skill acquisition

Objective: BodyMind Dancing^{SM}, Martha Eddy's form of contemporary dance based in Dynamic Embodiment^{SM}, a type of somatic movement education based in Laban/Bartenieff studies and Body-Mind Centering®, offers skills for mediating conflicts within dance. This workshop will demonstrate creative resolution of two typical conflicts in teaching or choreographing: 1) technique versus improvisation, and 2) beginning versus advanced dancers. The role of technique is common to all dance training whereas the use of improvisation varies. Examples of lay dancers and professionals collaborating in classes at international festivals and in universities as well in as in choreography such as Global Water Dances demonstrate how BodyMind Dancing has created an intersection point for learning dance that is equally satisfying for all levels of dancers, using improvisation and technique.

Methods: Participants will engage experientially in BodyMind Dancing in a full dance space for 20 minutes and then spend 40 minutes involved with laboratory style learning. We will use BodyMind Dancing's somatic awareness to access physiological resources common to all bodies that facilitate the reconciling of tensions between learning through technique and learning through improvisation. This is accomplished by sharing information sourced from the biological sciences – the different physiological systems: the bones, muscles, nerves, brain, glands, organs, soft tissues, and fluids. Next, using BodyMind Dancing's "BARRE" matrix, a template is provided for meeting the needs of both technical skill development (e.g., dynamic alignment, balance, coordination, movement memory, range of motion) and of personal and group expression (e.g., accessing emotions through diverse movement dynamics, communicating meaning, shaping body configurations, managing expectations). These ideas will be explored through both technical phrases and guided improvisations.

Results: In our time together, dancers, choreographers, dance educators and therapists can experience how Dynamic Embodiment's Embodied Physiology^{SM} approach to movement awareness sharpens dance skills amongst all levels of students, AND experience its relationship to enhanced technical mastery and improvisational movement expression.

Conclusion: This interactive presentation seeks to demonstrate how the scaffolding of both improvisation and technical skills can sort out two conflicts in dance – technique versus improvisation and the stress of working together with dancers with diverse technical abilities toward a common goal.

In: Movement 2019: Brain, Body and Cognition
Editors: Joav Merrick and Gerry Leisman

ISBN: 978-1-53616-714-6

Chapter 72

MEMBRANES AS CHANGE AGENTS: A DYNAMIC EMBODIMENT APPROACH TO EMBODYING OUR CELLS WHILE DANCING

Martha Eddy[*]
Dynamic Embodiment Somatic Movement Education and Therapy,
New York, United States of America

Keywords: dance, movement, education

Dynamic Embodiment (DE) is an ISMETA approved training of somatic movement education and therapy developed in 1990 that distinguishes itself from other somatic movement systems through its use of DANCE to train practitioners. As part of each intensive DE trainees engage in BodyMind Dancing to learn theoretical concepts ranging from biology, motor development, neuroscience and physiology as well as injury prevention and injury rehabilitation. Steeped in the art and science of qualitative movement observation, Dynamic Embodiment draws on both Laban Movement Analysis (LMA) research methods and Body-Mind Centering® (BMC)'s experiential anatomy and neuro-motor developmental theories. In this workshop we will focus on DE's use of LMA coding to better understand how BMC views the motor activity of cells through the expression of human movement.

Objective: The objective of membranes as change agents: A dynamic embodiment approach to embodying our cells while dancing is to provide participants with an embodied experience of cellular activity for two reasons 1) to learn more about the human body and the health of the dancing body and 2) to gain metaphors for dance choreography and pedagogy from cellular biology. In order to meet these goals the methods include discussion, visual projections, movement games and dance improvisations. The initial focus of the workshop will be on early developmental aspects of the cell, the parts and structure of cells, and the differing conditions of diverse cell types. Next we will spend time learning about the cell membrane and how it functions with permeability. The expressions of permeability will inform both dynamic changes in movement performance in force, time, space and tension as well as interactive experiences of fluid and rigid boundaries. The results of this workshop will include a group improvisation that could be (with participants permission) either filmed or demonstrated at some future point for Movementis.

Conclusion: In conclusion, the interactive framework of dance will serve to dig deeper into cellular biology and cellular exchange for the purposes of art-making and health enhancement.

In: Movement 2019: Brain, Body and Cognition
Editors: Joav Merrick and Gerry Leisman
ISBN: 978-1-53616-714-6

Chapter 73

DIVING INTO OUR FLUID BODY: THE MOVEMENT OF CONTINUUM

Batyah Schachter*
Jerusalem Academy of Music and Dance, Hebrew University,
Continuum Teachers Association, Jerusalem, Israel

Keywords: movement, sound, wave motion, attention

The exploration in this workshop will focus on noticing and exploring the effect of sound on the body's movement response. Bringing about a direct experience of the connection between sound, attention and movement we will explore the difference between creating a movement, being moved and being present in our body's sensations and movements. We will experiment with different sounds and their various effects, how they differ in different individuals and in changing body's conditions.

The tools of Continuum practice are extremely simple, using breath and sounding instructions followed by a period of allowing the flow of movement that emerges as a response. It does not require any prior experience as a mover or dancer, yet the experience can be noticeable to everyone willing to pay attention and follow themselves.

Continuum is a movement approach and method that was developed by the late Emilie Conrad in Los Angeles in the past 50 years. It is an exploration of the fluid nature of our body that enables us to dissolve into the movement that occurs within us continuously. Turning our attention to the ever present wave motion within, inevitably draws us into a different kind of focus and awareness. In Continuum we use our own sound in order to invoke the wave motion in the body, and then we let that awakened wave become the choreographer of our movement. When we turn our attention, openly and willingly to that motion, we find that movement is allowed to happen rather than being executed by our will's command.

Moving in this way we unravel deeply ingrained patterns and bypass the known possibilities of our body's movement. We turn our attention into a vastly different dimension of our body, which can bring about unexpected sensations and movement vocabulary

In: Movement 2019: Brain, Body and Cognition
Editors: Joav Merrick and Gerry Leisman
ISBN: 978-1-53616-714-6

Chapter 74

REAL TIME COMPOSITION

Anat Shamgar*
The Jerusalem Music and Dance Academy, Jerusalem, Israel

Keywords: dance, movement, composition

The content of the talk (conceptual presentation) will be the conflict within "Real time composition" as an improvisational dance composing. The focus is the constant process of clearing one's cognition. Clearing inhibition, pre-supposed notions, automatic gestures, memories, connotations, images, citations… Reaching a clear. While clearing the body-mind, one gives way to "reading" what is revealed: A new body of notions surfaces. Knowledge, memory, direction, logic, organization, structure, as well as emotions and sensation.

The process of "Real time composition" is a conflict in its core: While deconstructing pre-supposed knowledge – a new information surfaces from the cleared action: to the mind, in the body, into the space, and through the action.

"Real time composition" is a climax of a creation process, in the place where the body and the mind are in multi-level dialogue, and it evaporate into the process in the process. Alertness is at their climax, all resources are devoted to the dual process of clearing, and absorbs the content which lurks into the "vacuum" which the clearing process opens. Fills it with new, evaporating content, like a black hole in the mind. So, condensed that it sponges in all immediate creations.

In: Movement 2019: Brain, Body and Cognition
Editors: Joav Merrick and Gerry Leisman
ISBN: 978-1-53616-714-6

Chapter 75

NEUROMOTOR RESOURCES FOR INTEGRATING SECURE ATTACHMENT AND TRAUMA RECOVERY

Talia B Shafir*
Department of Somatic and Spiritual Psychology,
International University of Professional Studies,
Maui, Hawaii, United States of America

Keywords: somatic movement, neuromotor, trauma, attachment, PTSD

The objective of this workshop is to bring awareness, through movement, to the integrated continuum of physical and psychological development that affects us all throughout the lifespan. Based upon the parallel progression of muscular activation and psycho-social maturation, movement out into the world requires an internalized somatic as well as a cognitive base to support secure attachment. Healing from trauma and the aftermath of post-traumatic stress (PTSD) depends upon the status of an adult's attachment system in the face of events that overwhelm the nervous system.

Trauma victims are often caught in a freeze state of the polyvagal system's protective defense mechanism, making social re-integration appear as an insurmountable obstacle driven by unresolved attachment defense processes. Movement serves as a somatic lifeline to what cannot be reached through talking about the experience. Developing missing somatic resources, and connecting existing ones, helps to build resilience and expand consciousness.

Participants will work experientially with the connection between thoughts and movement patterns. We will practice using observational skills to identify and assess disruptive somatic patterns in adults that underlie chronic trauma/PTSD and interfere with healthy independence and stable relationship. Clinical emphasis will be on working with muscular tonicity and the use of developmental movement as support for interactive and self-regulation, as well as tracking changes in thought patterns.

This workshop will consist of a combination of movement (75%) and lecture (25%) on the connection between an adult's attachment status and the somatic and psychological resources available for dealing with trauma and its aftermath. The material is based upon Attachment Theory and Attachment Status as addressed by the Adult Attachment Picture System (AAP) as well as several somatic psychotherapeutic approaches to trauma, PTSD, and developmental disruption.

In: Movement 2019: Brain, Body and Cognition
Editors: Joav Merrick and Gerry Leisman
ISBN: 978-1-53616-714-6

Chapter 76

LEARNING WITH VR-BASED ASSISTIVE SYSTEMS

Thomas Schack*[*], *Cornelia Frank, Felix Hülsmann and Ludwig Vogel

Neurocognition and Action Research Group, Center of Excellence Cognitive Interaction Technology (CITEC), Faculty of Psychology and Sports Science, Bielefeld University, Bielefeld, Germany and Graphics and Geometry Research Group, Center of Excellence Cognitive Interaction Technology (CITEC), Bielefeld University, Bielefeld, Germany

Keywords: virtual reality, assistive systems, motor learning, mental representation

Objective: The understanding of the neurocognitive architecture of actions based on empirical research is an important step in applied fields such as the mental coaching of athletes in high-performance sports or rehabilitation. On the other hand, a fundamental aspect of the growing field of cognitive systems, particularly in relation to its central goal of action support, is to be able to identify user and context specific problems in the execution of action sequences. Based on such information intelligent assistive systems should then be able to provide adequate feedback, i.e., by displaying situation and user-specific help comments and hints.

Methods: We developed a mobile adaptive assistance system in the form of intelligent glasses (ADAMAAS-glasses) that provide unobtrusive, anticipative and intuitive support in every-day situations. The system is able to identify problems in ongoing action processes, react to mistakes and provide context-related assistance in textual, pictorial or avatar-based formats superimposed on a transparent virtual display. Furthermore in cooperation with an interdisciplinary team of researchers we used a multi-modal approach and developed a virtual sport coaching device, called Intelligent Coaching Space (ICSpace). Within the ICSpace teacher-trainee scenario, the virtual coach supervises the acquisition of action sequences by verbally instructing participants and assessing changes of mental representation during skill learning in order to provide feedback. We investigated learning processes with these two systems and we checked for instance how modeling one's own future performance during action observation and mental imagery impacts motivational, cognitive and motor variables. We measured movement quality, mental representation structure, and self-efficacy of a squat movement.

Results: Preliminary findings indicate that the group with future modeling profit more compared to the group with the modeling of recent motor performance in motivational (i.e., self-efficacy), cognitive (i.e., mental representation structure) as well as motor (i.e., movement quality) variables.

Conclusion: These findings suggest that simultaneous imagery whilst observing future states of action helps establish both cognitive and motivational prerequisites that enable better motor performance, and that virtual reality is a promising tool to create learning environments that exceed real-world opportunities.

In: Movement 2019: Brain, Body and Cognition
Editors: Joav Merrick and Gerry Leisman ISBN: 978-1-53616-714-6

Chapter 77

INDIVIDUAL COGNITIVE DIAGNOSTIC IN WORK PROCESSES FOR PEOPLE WITH COGNITIVE DISABILITIES

Ludwig Vogel* and Thomas Schack
Neurocognition and Action Research Group, Center of Excellence Cognitive Interaction Technology (CITEC), Faculty of Psychology and Sports Science, Bielefeld University, Bielefeld, Germany

Keywords: SDA-M, cognitive representation, cognitive diagnostic

Objective: Mental representations built the cognitive basis in the organization and control of actions. To act meaningful and goal-oriented, actions and their representation should be organized task-depended in a functional manner. However, oftentimes people with cognitive disabilities have problems to fulfill working-tasks autonomous and self-organized. In this study, we investigated the cognitive representation of complex actions in work processes in trainees with cognitive disabilities and compared them to their instructors.

Methods: Therefore, we applied a psychometric measurement method. The structure dimensional analysis of mental representation (SDA-M) is procedure that can analysis the cognitive relationship of work tasks. The trainees were mentally handicapped employees in a vocational training program that prepares handicapped individuals to participate in workplaces. We investigated the work process in a cafeteria.

Results: We found significant differences in the cognitive representations of the work processes between the instructors and the trainees. Instructors showed task-specific cognitive structure with well-structured representation of task specific actions, whereas trainees exposed less functional structure representations.

Conclusion: Based on these results we will present solutions for individual cognitive training. With this study we could show that it is possible to measure the cognitive structure of a work process in people with cognitive disabilities and provide individual feedback for learning processes.

In: Movement 2019: Brain, Body and Cognition
Editors: Joav Merrick and Gerry Leisman
ISBN: 978-1-53616-714-6

Chapter 78

ON THE PRENATAL COGNITIVE LIFE OF THE FETUS IN ACTION

Gerry Leisman*

The Faculty of Health Sciences, University of Haifa, Haifa, Israel, The National Institute for Brain and Rehabilitation Sciences, Nazareth Israel and Department of Clinical Neurophysiology, Institute for Neurology and Neurosurgery, University of the Medical Sciences, Havana, Cuba

Keywords: fetus, movement, cognition, learning, sensory-motor interaction, development

The human brain is a complex organ designed to effect cognitive functions, such as attention, perception, memory, learning, and action. All of the nervous system is active during prenatal development and the aim of the presentation is to present data to allow us to better understand fetal cognitive functions. Fetal cognitive motor activity, including auditory discrimination, orienting, the wake–sleep cycle, FHRs, and defensive reactions, appear to be under the reflexive control of the brainstem, which also appears capable of learning-related activity. However, the fetus begins to processes sensory stimuli at a cortical level, including pain, from approximately 25 weeks of gestation. At gestational week 34, the fetus not only perceives complex external acoustics but can also discriminate between different sounds. Fetal action planning and anticipation of movement and action, a sense of agency, is established by 22 weeks and investigations using four-dimensional ultrasound demonstrate that complexity of fetal motor action and behavior increases as pregnancy progresses. The capacity of the fetus to learn and memorize are significant. At term, subcortical structures of the brain are well developed. There is high activity in primary cortical areas and low activity in association areas.

Clinically relevant data on cognitive functions of the fetus could be important for the management of fetal pain and treatment of preterm infants as well as for improved neurodevelopmental outcome of fetuses from high-risk pregnancies. Finally, the brain's developmental journey, including development of cognitive functions, continues with the same intensity in the postnatal period.

In: Movement 2019: Brain, Body and Cognition
Editors: Joav Merrick and Gerry Leisman
ISBN: 978-1-53616-714-6

Chapter 79

MOVING THE MIND: EMBODIED COGNITION IN PSYCHOTHERAPY

Tania Pietrzak*

Olga Tennison Autism Research Centre, School of Psychology and Public Health,
La Trobe University, Melbourne, Victoria, Australia

Keywords: embodied cognition, psychotherapy, embodiment techniques, emotional regulation

Aim: This workshop aims to broaden the concept of cognition to include new concepts of embodiment in clinical psychotherapeutic application. Embodiment sciences are pointing to a broadened concept of cognition that is anchored in perception and actions of the body. Mind–body reciprocity means that the mind is embodied and the body is mental. The body is a co-designer of mental processes and is pivotal for embodied interaction. The theoretical position behind the workshop is that the body builds the mind and therefore must be attended to in psychotherapy.

Embodiment interventions can enrich traditional psychotherapy approaches by giving clients a deeper and more accelerated understanding of their problems and emotions by using the body's display of important information from the unconscious impulsive system. These somatic markers can be translated into verbal language using pictorial language to develop a nuanced self-awareness in the client. The workshop is intended to give participants practical hands-on up-to-date instructions on how to apply embodiment techniques in psychotherapy.

Specific Objectives: Participants will gain: a) A theoretical understanding of embodiment drawing from dual systems theory, b) Knowledge of empirical research on the embodied emotions, c) Understanding that problematic situations are connected with a network of emotions and affect laden fragments of learning history, d) Empirically grounded embodiment techniques and how to apply them in clinical settings to help clients regulate their emotions, e) How to use the Emotional Field to elicit clients' primary and secondary emotions, f) Knowledge on how clients' primary and secondary emotions are linked to contradictory action impulses and conflicting schemata and g) How to resource clients with adaptive ways of using all their emotions to reach their values driven goals and to develop a new life strategy.

Workshop methods: a) Psychoeducation, b) a tool kit (manual) of strategies to be applied in clinical application, c) live and video demonstration showing the method and d) self-experience in small groups.

In: Movement 2019: Brain, Body and Cognition
Editors: Joav Merrick and Gerry Leisman

ISBN: 978-1-53616-714-6

Chapter 80

AUGMENTING AROUSAL BASED NEUROFEEDBACK WITH BIO-PLAYBACK TO IMPROVE GENERALIZING SKILLS IN ASD

Lynette Louise*
Independent Practice, Brain and Body, Simi Valley, California, United States of America

Keywords: autism, play/behavior therapy, neurofeedback, augmentation, movement

Objective: Detail the manner in which combining neurofeedback with biofeedback influenced play can improve the ability to generalize skills for the autistic child. Autism spectrum disorders (ASD) are neurodevelopmental disorders with a wide range of functionality, including, but not limited to, significant challenges in communication, social skills and behavior. The impact this disorder has on our global community is vast. The Center for Disease Control (CDC) estimated the number of eight year old ASD children in America (ranking fifth out of ten countries surveyed by the statistics portal) rose from 1/150 to 1/59 in just 14 years (2000 to 2014). Applied Behavior Analysis (ABA) is the primary behavioral/teaching approach used internationally. Unfortunately, ABA suppresses the child's desires and increases already elevated stress levels. Conversely, play therapy engages the child, teaching skills while immersed in movement. Neurofeedback is analogous of a type of play therapy teaching tool for brain waves. Combining the two improves outcomes.

Methods: Neurofeedback is commonly referred to as EEG biofeedback. Biofeedback is a method of using technology to immediately gain access to physiological information that is otherwise unknowable. When brain waves are corrected utilizing neurofeedback, improved brain function is the result. Of course, a better functioning brain still needs the correct information for desirable behavior to occur. ASD individuals are mystified by their effect on the environment. To correct this challenge it is important they learn in the moment. Bio-playback is a nonjudgmental person-delivered immediate information response technique meant to be augmented by neurofeedback for faster results.

Results: This method has been utilized in homes, schools and care units in countries and cultures as diverse as Uganda, Israel, Switzerland and America. Desirable change is seen in behavior and generalized into other environments within days. Brain wave behavior is also improved as evidenced by pre and post EEG samples with protocol explanations.

Conclusion: Inability to generalize skills is not just an inherent problem in ASD but also in our method of teaching. Gaining widespread understanding of these techniques will ultimately reduce isolation and increase generalization of skills in the ASD community and beyond.

In: Movement 2019: Brain, Body and Cognition
Editors: Joav Merrick and Gerry Leisman
ISBN: 978-1-53616-714-6

Chapter 81

THE INFLUENCE OF OCCUPATIONAL THERAPY ACTIVITIES ON DAILY LIVING AND QUALITY OF LIFE OF CHILDREN HAVING ACUTE LYMPHOBLASTIC LEUKEMIA

Justinas Blaževicius*[*], *Katrin Kocan and Aurelija Šidlauskiene
Department of Rehabilitation, Physical and Sports Medicine, Health Science Institute, Faculty of Medicine, Vilnius University, Lithuania

Keywords: acute lymphoblastic leukemia, occupational therapy, quality of life, daily activities, children

Objectives: Assess the influence of occupational therapy activities for the children's daily activities and living quality, being ill with acute lymphoblastic leukemia.

Methods: 54 patients, ill with acute lymphoblastic leukemia, participated in the investigation: 51.85 per cent boys and 48.15 per cent girls. The investigation was conducted in a questionnaire survey manner from September of 2013 to March of 2015. The investigating patients were examined twice: at the start and end of the occupational therapy activities. During the investigation, the PedsQL pain scale, PedsQL living quality questionnaire, COPM, HAD (depression and dismay assessment), and dynamometry readings were used.

Outcomes: The children at the age of 5-18 years assessed their physical health inferior than their psycho-social health. The principal of children's dismay and causes of their complaints: fear of procedures (31.79 ± 31.97), pain experienced during the treatment (54.40 ± 24.18), communication with the medical staff (55.40 ± 18.96) and nausea during the treatment. Statistically significant improvement was identified for the children, being ill with ALL, across all the age groups after applying the procedures of occupational therapy activities. The children assessed their quality of life improved. Assessing the shifts in the performance and satisfaction of activities of the children, ill with ALL diseases before and after occupational therapy activities, using the Canadian activity assessment test, it was identified that the occupational therapy had a statistically significant impact upon the performance of daily activities ($p = 0.000 < 0.05$), its impact was not equal to all activities ($p = 0.000 < 0.05$), but equal to all the age groups ($p = 0.194>0.05$) and the impact of occupational activities to the different age groups and activities did not differ significantly and statistically ($p = 0.377>0.05$).

Conclusion: Occupational therapy activities had a statistically significant impact upon the children's dynamometry results ($p = 0.000 < 0.05$). This impact was different among the age groups ($p = 0.000 < 0.05$). It was identified that the dynamometry shift in the age group of 5-7 years was statistically significant ($p = 0.046 < 0.05$) in relation to the quality of life shift, this relationship is positive (larger dynamometry shift promotes a larger life quality shift), in all the other cases, the dynamometry shift did not have any impact upon the life quality nor in the assessment of the daily activities.

In: Movement 2019: Brain, Body and Cognition
Editors: Joav Merrick and Gerry Leisman
ISBN: 978-1-53616-714-6

Chapter 82

SCHOOL-AGE CHILDREN LIFE PECULIARITIES DURING ONCOHEMATOLOGICAL DISEASE

Laura Petkunaite, Justinas Blaževicius* and Katrin Kocan
Department of Rehabilitation, Physical and Sports Medicine, Health Science Institute, Faculty of Medicine, Vilnius University, Lithuania

Keywords: oncohematological disease, children, psychoemotional condition, self-sufficiency, isometric hand muscle grabbing strength, sense of psychical pain

Objectives: To study the peculiarities of onco-hematological disease for the school-age child's psycho-emotional state, self-sufficiency, sense of physical pain, isometric hand muscle grabbing strength and leisure time.

Methods: The study was done in 2017 - 2018, VUL SK branch, children hospital, the division of onco-hematology. Subjects were 20 schoolchildren. Hospital Anxiety and Depression Scale, Functional Independence Measure, dynamometer, Visual Analogue Scale and Canadian Occupational Performance Measure were used for research.

Results: Anxiety ($p = 0,929$) and depression ($p = 0,259$) satisfies the normal mood of the subjects, Hospital Anxiety and Depression Scale amount of points is bigger than 7 scores ($p < a$, $p = 0.022$). When comparing the healthy population and the muscle strength of the test, the difference between the right hand of the boys is 13 kg and the left one is 11.8 kg, between girls is 7.8 kg and 7.6 kg. Functional Independence Measure average is different from the norm ($p < 0,001$). The average of sense of physical pain – 25,2 mm. The performance of leisure activities is estimated at 5.8 points. As the performance increases, satisfaction with the activities is increasing (rho = 0,961).

Conclusion: More than half of schoolchildren with onco-hematological diseases (55%) have a psycho-emotional disorder ($p = 0.022$). Boys are mostly moderately affected, and girls' mood fits a set norm. No statistically significant changes were observed in the individual assessment of anxiety and depression symptoms. The average of isometric hand muscle grabbing strength in patients with onco-hematological diseases, taking into account the age and gender of the subjects, is lower than that of the healthy population. A bigger difference is noticeable in the boy's population. The degree of independence of the schoolchildren with onco-hematological diseases ranges from the average need for help to full self-sufficiency. In most cases, children do not need support at the time of contact, they are almost independent, more time is needed for their activities, or auxiliary measures are used. On average, these children experience mild physical pain during inpatient treatment. Leisure time in schoolchildren with onco-hematological disease becomes more passive than usual due to the emergence of symptoms and side effects of the treatment. The activities performed during the leisure time, and satisfaction is assessed on average by the subjects, mostly due to fatigue.

In: Movement 2019: Brain, Body and Cognition
Editors: Joav Merrick and Gerry Leisman
ISBN: 978-1-53616-714-6

Chapter 83

TAI CHI INTERVENTION PROGRAM FOR IMPROVING POSTURAL STABILITY AND GAIT IN PEOPLE WITH PARKINSON'S DISEASE: A STUDY PROTOCOL

Nok-Yeung Law [*] ***and Jing Xian Li***
University of Ottawa, School of Human Kinetics, Ottawa Canada

Keywords: Tai Chi, Parkinson's disease, postural stability, gait

Background: Parkinson's disease (PD) is a progressive, neurological condition that adversely affects gait, postural stability and cognitive capacity. Regular physical activity can improve gait and posture as well as cognitive function by providing a neuroprotective and neurorestorative effect on the brain within this population. However, exercise intervention approaches that considers both aspects of improving movement capacity and cognitive function are still lacking.

Objective: The study's aim is to develop a biomechanical based tai chi intervention program for individuals with early stage Parkinson's disease (PD) and to examine its effects towards improving postural stability and gait in this population.

Method: A biomechanical based tai chi program will be developed for participants with early stage PD by analyzing the kinematic, kinetic, and neuromuscular activities of seven selected tai chi movement as performed by an experienced tai chi practitioner. The seven tai chi movements emphasize movement in medial-lateral direction and have a high demand in the muscle activity of trunk and lower limb, particularly the hip abductors that are closely related to postural stability. A 12-week tai chi intervention program will be used to examine whether gait, postural stability, and cognitive function can be improved among a group of participants with early stage PD (Hoehn and Yahr stages 1 to 3) and will be compared to a control group of participants with PD who will be asked to maintain their regular physical activity levels. A total of 40 participants will be recruited for the study and these participants will randomly be assigned to either the tai chi or control groups. Participants in the tai chi group will receive tai chi training while the control group will be allowed to engage in regular physical activity; each group will participate in 60-minutes of tai chi training or physical activity per day, three times a week for 12 weeks. The gait and postural stability in both groups will be evaluated through biomechanical analysis of obstacle crossing at baseline, 6-week, and 12-week after the start of intervention period. The cognitive performance will be measured by scores on the Stroop Test, Wisconsin Card Sorting Test, and Trail Making Test (Part A and B) that relate to executive functioning.

In: Movement 2019: Brain, Body and Cognition
Editors: Joav Merrick and Gerry Leisman
ISBN: 978-1-53616-714-6

Chapter 84

VISUAL CUES DOMINATE ACTION-PLANNING WHILE SENSORIMOTOR PROCESSES DOMINATE PERCEPTION IN OBJECT MANIPULATION IN THE YOUNG AND ELDERLY

***Thomas Rudolf Schneider*[1,*], *Gavin Buckingham*[2] and *Joachim Hermsdörfer*[1]**
[1]Human Movement Science, Department of Sport and Health Sciences, Technical University of Munich, Munich, Germany
[2]Sport and Health Sciences, College of Life and Environmental Sciences, University of Exeter, Exeter, United Kingdom

Keywords: object manipulation, internal models, torque perception, weight perception, motor error

Objective: We investigated the interplay of geometric visual processing, sensorimotor prediction errors and initial guesses in anticipatory torque control and the perception of torques and weight in an object manipulation task.

Methods: Young and elderly subjects repeatedly lifted an inverted-T-shaped object in which the center of mass was randomly varied between trials by repositioning a hidden weight in the first experiment (no visual cues) and additionally randomly repositioning the handle position in experiment 2, providing salient geometric cues on the weight distribution. With force-torque sensors mounted below the handle-surfaces, we analyzed the finger-positioning and total torque exertion at the moment of lift-onset, determined by a position/orientation-sensor, as an indicator of predictive torque control. Subjects indicated torque guesses and percepts with a digital caliper and weight percepts by oral rating. We employed mixed-effects multiple regression modelling.

Results: In the absence of cues, we found that participants partly adhered to committed planning errors in addition to adapting to previous trial's torque. Interestingly, planning errors lead to an enhanced torque perception. Furthermore, both planning errors and the actual torque induced an illusionary percept of increased heaviness. Given geometric cues, these governed torque anticipations, rendering sensorimotor influences secondary. The perception of torques and object weight was however still dominated by the impact of the actual torque and the committed planning error, while geometric cues slightly modulated the torque percept in the elderly and curbed heaviness percepts in both groups. Besides, subject's torque judgments were influenced by their initial guesses. In both experiments torque perception did not influence upcoming motor control. We found no major differences between the age groups.

Conclusion: The CNS employs feedforward predictive models to refine both upcoming motor predictions and shape the percept of object properties. In the absence of cues on the object properties, the motor system partly retained previous sensorimotor memory states and partly adapted them towards experienced errors, while the perceptual system emphasized sensorimotor feedback of prediction errors. Given geometric cues, the CNS linearly transformed these to adequate motor plans and suppressed sensorimotor influences while relying on sensorimotor feedback about object dynamics and prediction errors in object perception.

In: Movement 2019: Brain, Body and Cognition
Editors: Joav Merrick and Gerry Leisman
ISBN: 978-1-53616-714-6

Chapter 85

DIFFERENT MYOELECTRIC PATTERNS PREDICT MANUAL REPETITIVE TASK DIFFICULTY AND ENDURANCE IN YOUNG AND OLD ADULTS

Maxana Weiss*, Christopher A Bailey and Julie N Côté
Department of Kinesiology and Physical Education, McGill University, Montreal, Quebec, Canada and Centre for Interdisciplinary Research in Rehabilitation (CRIR site), Jewish Rehabilitation Hospital Research Center, Laval, Quebec, Canada

Keywords: aging, electromyography, fatigue, repetitive work

Objective: Identify the influence of old age on myoelectric manifestations during a repetitive upper limb task, and their association to perceived task difficulty and endurance.

Methods: Healthy young (YOUNG; N = 18, 11 women) and old (OLD; N = 12, 10 women) adults were outfitted with surface electromyographic (sEMG) sensors, and reference contractions were recorded from six upper limb muscles on the dominant side: Upper (UT), Middle (MD), and Lower (LT) Trapezius; Anterior (AD), Middle (MD), and Posterior (PD) Deltoid. Participants performed a repetitive tapping motion between two points placed on a table at 30% and 100% of functional reach, from a seated position, following the beat of a metronome. The task continued until participants reached a rate of perceived exertion on the BORG-CR 10 scale = 8 or 45 minutes duration. EMG amplitude was normalized to reference contractions, and root-mean-square (RMS), Median Power Frequency (MdPF) and RMS Coefficient of Variation (COV) were averaged over the middle ten forward movements of the first task minute (Min1).

Correlation coefficients were run between Min1 EMG measures with the times it took participants to reach BORG-CR = 5 “Hard” (B5) and to reach fatigue (FT).

Results: There was no group difference in the time to B5; however OLD reached FT earlier than YOUNG. There were group-specific correlations between Min1 EMG patterns and times to both B5 and FT. In YOUNG, Min1 EMG RMS of MD ($r2 = 0.550$; $p < 0.05$), MT ($r2 = 0.519$; $p < 0.05$), and LT ($r2 = 0.583$; $p < 0.05$) were positively correlated with B5, and MdPF MT ($r2 = -0.644$; $p < 0.01$) was negatively correlated to FT. In OLD, UT EMG RMS ($r2 = -0.744$; $p < 0.01$) was negatively correlated to FT whereas PD COV ($r2 = 0.562$; $p < 0.05$) was positively correlated to FT.

Conclusion: Results suggest that in young adults, engaging the scapula stabilizers when starting a repetitive arm task helps delay the perception of this task being “Hard”. In older adults, those who initially relax their upper trapezius and employ variable deltoid activation patterns perform this task longer. Results can help better understand how old adults can fare safely and efficiently in work engaging manual repetitive work.

In: Movement 2019: Brain, Body and Cognition
Editors: Joav Merrick and Gerry Leisman
ISBN: 978-1-53616-714-6

Chapter 86

Exploration in the Presence of Mother in Typically and Non-Typically Developing Pre-Walking Human Infants

Tzviel Frostig[1], Hanna Alonim[2], Giora Scheingesicht[2], Yoav Benjamini[1] and Ilan Golani[3,*]
[1]Department of Statistics and Operation Research, School of Mathematics and Sagol School for Neurosciences, Tel Aviv University,
[2]Mifne Center for Early Treatment of Autism, Rosh Pinna
[3]School of Zoology and Sagol School for Neurosciences, Tel Aviv University, Tel Aviv, Israel

Keywords: origin-related exploration, homebase, autism, ASD, kinematics

Objective: Comparison of exploration of human pre-walking typically-developing (TD) and non-typically developing (NTD) infants in mother-infant dyads.

Methods: The NTD mother-infant dyads had been referred to a center for the treatment of autism by pediatric neurologists and expert clinicians. We used an arsenal of computational tools previously developed for the study of origin-related exploration in animals, including video filming, tracking, smoothing, segmentation, and algorithmic analysis.

Results: We documented in the TD mother-infant dyads a phylogenetic ancient behavior: origin-related exploration. Strikingly, while the TD infants exhibited excursions in reference to mother and close physical proximity with mother when visiting her, the NTD infants tended to avoid mother's place, performing few if any excursions, and exhibiting poor physical proximity with mother. Significant statistical differences were found between the following TD and NTD infants' behavioral endpoints: number of excursions performed from mother, time near mother, speed, percentage of room covered, and number and duration of contact with mother episodes.

Conclusion: Our study yielded a candidate referral tool for pediatric neurologists. Given the pervasiveness of origin-related exploration in invertebrates, vertebrates, and primates, we now face a challenge to find an animal model that will exhibit active investigation (perhaps involving focused attention to restricted places or objects) while ignoring or suppressing the return to the origin, be it a mother or any other reference location.

In: Movement 2019: Brain, Body and Cognition
Editors: Joav Merrick and Gerry Leisman
ISBN: 978-1-53616-714-6

Chapter 87

VALIDITY AND INTER-OBSERVER RELIABILITY OF THE TURN 180 TEST TO IDENTIFY OLDER ADULTS WHO REPORTED FALLS

***Yulia Gamerman*[1,*], *Moshe Hoshen*[2], *Avner H Cohen*[3,4], *Zhana Alter*[5], *Luzit Hadad*[5] *and Itshak Melzer*[1]**

[1]Schwartz Rehabilitation and Movement Analysis Laboratory, Physical Therapy Department, Recanati School for Community Health Professions, Faculty of Health Sciences, Ben-Gurion University of the Negev, Beer-Sheva,
[2]Clalit Research Institute, Chief Physician's Office, Clalit Health Services, Tel Aviv,
[3]Sackler Faculty of Medicine, Tel Aviv University, Tel Aviv,
[4]Pediatric Ambulatory Center, General Health Services, Petach Tiqva
[5]Rotschild Physiotherapy Department, Clalit Health Services, Petach Tiqva, Israel

Keywords: TURN180 test, balance, fall risk, aging, validity, reliability

Background: Falls while turning are associated with increased risk of hip fracture in older adults. Reliable and clinically valid methods of turn ability assessments are needed.

Objectives: To explore the inter-observer reliability and known group validity of the TURN 180 test.

Methods: Seventy-eight independent old adults (mean age of 76.6 ± 6.5 years), were divided into three groups: Non-Fallers, fallers who reported 1-2 falls a year, and recurrent fallers who reported more than two falls a year. Participants underwent performance-based tests: Timed "Up and Go" test (TUG), Performance Oriented Mobility Assessment (POMA), and Berg Balance Scale Test (BBS). TUG was videotaped for later analysis of the TURN 180 test by two blinded observers.

Results: A significant difference were found in the TURN 180 test parameters between the groups ($p < 0.04$). The TURN 180 test is highly correlated with TUG ($r = 0.81-0.89$, $p < 0.001$) and BBS ($r = -0.704-0.754$, $p < 0.0001$), and moderately with POMA ($r = -0.641-0.698$, $p < 0.0001$). The number of steps was found to be the strongest parameter to determine fallers among older adults (specificity 96.3% and sensitivity 40%). Inter-rater reliability (ICC 0.91-0.96, $p < 0.0001$) was found to be excellent for the number of steps, time taken to accomplish a turn and total test score categories.

Conclusion: The data demonstrates that the TURN 180 test is highly reliable and can identify the older adults that fall. This indicates that the TURN 180 test can serve as a good performance based examination for research or clinical setting.

In: Movement 2019: Brain, Body and Cognition
Editors: Joav Merrick and Gerry Leisman
ISBN: 978-1-53616-714-6

Chapter 88

A MULTI-LEVEL MODEL FOR UNDERSTANDING THE FACTORS PREDICTING HEALTH BEHAVIORS: PHYSICAL ACTIVITY AND HEALTHY NUTRITION HABITS AMONG STUDENTS IN ISRAEL

Hila Beck*[*], *Riki Tesler, Daniel Moran, Tatyana Kolobov and Yossi Harel-Fisch

Department of Health Systems Management, Faculty of Health Sciences, Ariel University, Ariel and The International Research Program on Adolescent Well-Being and Health, School of Education, Bar-Ilan University, Ramat Gan, Israel

Keywords: school principals, youth, health promotion policies, physical activity, nutrition

Objective: The World Health Organization has defined school systems as the significant framework for health promotion, because they provide a platform that is easily accessible to most children. The goal of the present study is to examine the association between health promotion policies of school principals and student-level factors that predict health behavior patterns: physical activity and proper nutrition.

Methods: This cross-sectional study is based on data from the Health Behavior in School-Aged Children (HBSC) survey that included 7,000 students in grades 6, 8, and 10 in secular and religious public schools and in Arab schools, and about 126 principals.

Results: The findings of the study indicate: at the student level boys perform physical activity more frequently than girls ($B = 0.80$, $p < 0.01$); with increasing age there is a decrease in the frequency of physical activity ($B = -0.34$, $p < 0.01$); students from a high socio-economic background are more physically active ($B = 0.10$, $p < 0.01$); and students in religious public schools are less physically active compared to students in secular public schools ($B = -0.37$, $p < 0.01$).

At the multi-level, implementation of a policy that encourages physical activity is a predictor of physical activity habits at the student level ($B = 0.09$, $p < 0.05$). Moreover, findings show that school nutrition policies ($OR = 0.89$, $p < 0.05$) and principals' commitment to health promotion ($OR = 1.13$, $p < 0.01$) predict healthy eating habits among students. Findings also indicate that students in the Arab sector consume less healthy food compared to students in public schools ($OR = 0.54$, $p < 0.05$).

Conclusion: The results of the current study emphasize the importance of implementing health promotion policies in schools by principals to improve students' physical activity and healthy eating habits.

In: Movement 2019: Brain, Body and Cognition
Editors: Joav Merrick and Gerry Leisman
ISBN: 978-1-53616-714-6

Chapter 89

TARGETED TRANSCRANIAL ELECTRIC STIMULATION TO MITIGATE THE DUAL-TASK COST TO GAIT SPEED IN OLDER ADULTS

Brad Manor[1], Junhong Zhou[1], On-Ye Lo[1], Alexa Ludington[1], Racheli Katz[2], Marina Brozgol[2], Pablo Cornejo Thumm[2], Giulio Ruffin[3] and Jeffrey M Hausdorff[2,4,5,*]
[1]Hinda and Arthur Marcus Institute for Aging Research, Boston and Harvard Medical School, Boston, Massachusetts, United States of America,
[2]Center for the Study of Movement, Cognition and Mobility, Neurological Institute, Tel Aviv Sourasky Medical Center, Tel Aviv, Israel,
[3]Neuroelectrics Corporation, Cambridge, Massachusetts, United States of America,
[4]Department of Physical Therapy and Sagol School of Neuroscience, Tel Aviv University, Tel Aviv, Israel
[5]Rush Alzheimer's Disease Center and Department of Orthopaedic Surgery, Rush University Medical Center, Chicago, Illinois, United States of America

Keywords: gait, dual-tasking, cognition, non-invasive brain stimulation

Objective: In older adults, dual-task walking activates brain networks related to both cognitive and motor function. We thus hypothesized that multichannel transcranial direct current stimulation (tDCS) that simultaneously targets primary cognitive and motor regions would reduce the dual-task cost to gait speed more than tDCS targeting either region separately or sham stimulation.

Methods: A multi-site, randomized controlled within-subject cross-over study was completed in 48 older adults free of overt illness or disease (mean ± SD age = 75 ± 6 years, 35 women). Participants completed four study visits, each separated by at least 72 hours, during which dual-task gait was assessed before and after 20 minutes of tDCS targeting 1) the left dlPFC and M1, 2) left dlPFC only, 3) M1 only, or 4) neither region (sham). Repeated-measures ANOVAs adjusted for age and sex tested the effects of tDCS condition, time (pre/post stimulation), and their interaction, on the dual-task cost to gait speed (i.e., percent change from single to dual-tasking), as well as gait speed within single- and dual-task conditions separately.

Results: Participants were successfully blinded to tDCS condition and reported no unexpected tDCS side effects. An interaction between condition and time was observed for the dual-task cost to gait speed ($p = 0.04$). This cost was smaller (i.e., better and closer to zero) following tDCS targeting the left dlPFC and M1 simultaneously, as well as the left dlPFC alone, compared to all other time points. A trend towards a similar condition by time interaction was observed for gait speed within the dual-task condition (i.e., gait speed appeared faster only following multi-target and left dlPFC tDCS, $p = 0.09$), but not the single task condition ($p = 0.78$).

Conclusion: tDCS targeting the left dlPFC—or the left dlPFC and M1—decreased the dual-task cost to gait speed in a relatively large sample of older adults, more than seen with M1 only or sham stimulation. Future studies are warranted to examine the duration of tDCS-induced benefits, as well as the mechanisms through which stimulating the left dlPFC gives rise to improved dual-task performance.

In: Movement 2019: Brain, Body and Cognition
Editors: Joav Merrick and Gerry Leisman
ISBN: 978-1-53616-714-6

Chapter 90

EVOLUTION OF MOTOR CORTEX EXCITABILITY THROUGH UNILATERAL ARM IMMOBILIZATION OVER NINE HOURS

Leonardo Furlan[1], Adriana Conforto[2,3] and Annette Sterr[1,*]
[1]School of Psychology, University of Surrey, United Kingdom,
[2]Das Clinicas, Department of Neurology, University of Sao Paulo, Sao Paulo
[3]Albert Einstein Hospital, Sao Paulo, Brazil

Keywords: motor cortex, immobilization, movement, neural plasticity

Temporary upper limb immobilisation (ULI) has been used to provide a model of induced short-term neural plasticity to examine the effects of motor learning-based rehabilitation interventions. More specifically, previous research has shown that ULI decreases the excitability of the cortical motor representations (CMR) and hence creates a situation of reduced sensorimotor experience or disuse common in hemiparesis. Reversing ULI-induced CMR changes through motor learning interventions, such as movement observation or motor imagery therefore holds important insights for treatment development, in particular with regards to low-functioning hemiparesis.

Studies on ULI have used different immobilization periods, ranging from hours to days. Knowing the length of time of immobilisation that is necessary to induce cortical depression is critical to determine the true feasibility of an immobilisation-based experimental paradigm as a test ground for rehabilitation interventions. The present study therefore examined ULI-induced changes over a period of 9 hours using neuronavigated TMS to map the right FDI immediately before, after 3, 6, and 9h of right ULI in a repeated measures design. 12 participants aged 23.4 ± 4.7 years completed the ULI protocol (ULI group). For statistical analysis, the test-retest reliability of the TMS procedure was obtained from a control group with no immobilisation (N = 9 aged 21.1 ± 1.8) to calculate the standard error of measurement and minimal detectable change (MDC95).

Test-retest reliability of the TMS mapping procedure, obtained from the control group was moderate, suggesting that the TMS procedure used in the present study provides a stable measure. In both groups, CMR measures at 3, 6, and 9h were not significantly different from baseline, respectively. Numerical variations across the three time points were observed for the immobilization group but these changes fell within the range of random measurement error. Participant-level data corroborated this null result, with all CMR change values smaller than the absolute MDC95.

The absence of an immobilization effect is at odds with studies showing significant decreases in excitability after a few hours of immobilization. We postulate that the present finding may be explained by the stimulation of the motor cortex induced by the TMS pulses throughout the immobilization period.

In: Movement 2019: Brain, Body and Cognition
Editors: Joav Merrick and Gerry Leisman
ISBN: 978-1-53616-714-6

Chapter 91

EMERGENCE OF GESTURAL SYMBOLS AND CONVERSATIONAL PROTOCOLS IN GESTURAL DIALOGUE EXPERIMENTS

Hiromi Oda[*]
Department of Information Management, Tokyo Online University, Tokyo, Japan

Keywords: gestural dialogue, emergence of symbols, body movements and cognition, communicative dynamism

Objective: The primary objective of the research project is to observe the processes in which participants with no previous exposures to sign languages adjust to the environment in which they can communicate only in gestures, and create a new "language." The basic results of individual experiments were already reported elsewhere. The objective of current analysis is to find out the characteristics of the symbols emerged in the experiments and conversational protocols, or grammar-like rules in gestural interactions.

Methods: The basic format of gestural dialogue experiment:

- Two participants sit at a table facing each other.
- There is a low partition so that they cannot see the partner's cards.
- At the beginning, only "yes" and "no" signs were agreed. Other signs must be agreed on the spot by the participants.

Materials: Three different formats of experiments:

- Basic format: For each set, six picture cards of simple objects, such as animals, people or plants. Participants must communicate the sequential orders of cards only using gestures. Six Japanese pairs x five sets.
- Location format: Communicate the locations of six objects in the common backgrounds. Six Japanese pairs x five sets.
- Movement format: Communicate the movements of two objects (the origin, path, and destination) in the common backgrounds. Six Japanese pairs x four sets.

Results: Average success rates for experiments were 91%, 68% and 70%, respectively. Successful participant pairs quickly agreed on the signs they use for the objects and developed protocols how they communicate locations and movements of objects. The symbols first emerge as imitative and descriptive gestured developed into more compact, short and quick signs as the sessions progressed. Some symbols used by participants were very similar to the symbols used in actual sign languages in other parts of the world. They also developed strategies to confirm that they were talking about the same thing using synchronized gestures, and to align the viewpoints they were taking while indicating "right side" or "left side."

Conclusion: Through these experiments, we can observe the cognitive processes in which gestural signs develop from iconic gestures to linguistic signs, and how participants agree on the gestural protocols to conduct successful gestural conversations.

In: Movement 2019: Brain, Body and Cognition
Editors: Joav Merrick and Gerry Leisman
ISBN: 978-1-53616-714-6

Chapter 92

CARDIORESPIRATORY FITNESS IS ASSOCIATED WITH HIPPOCAMPAL-RELATED COGNITIVE RESERVE IN OLDER ADULTS WITH AMNESTIC MILD COGNITIVE IMPAIRMENT

Tamir Eisenstein[1,2,*], Galit Yogev-Seligmann[1,2], Elissa Ash[3], Nir Giladi[2,3], Haggai Sharon[1,2], Talma Hendler[1,2] and Yulia Lerner[1,2]

[1]Sagol Brain Institute, Tel-Aviv Sourasky Medical Center, Tel-Aviv,
[2]Sackler Faculty of Medicine, Tel-Aviv University, Tel-Aviv
[3]Department of Neurology, Tel-Aviv Sourasky Medical Center, Tel-Aviv, Israel

Keywords: mild cognitive impairment, cardiorespiratory fitness, hippocampus, neuroimaging, memory

Objective: Hippocampal deterioration is apparent already at prodromal stages of Alzheimer's disease, namely amnestic mild cognitive impairment (aMCI). Increased hippocampal atrophy and dysfunction have been associated with poorer cognitive and clinical status in aMCI. Cognitive reserve (CR) refers to the variability between brain pathology and expression of clinical symptoms among patients. High levels of CR are associated with lifestyle factors including education, and leisure activities, and were suggested to mediate preserved cognition in the presence of advanced neuropathology by compensatory recruitment of neural resources in clinical populations. Cardiorespiratory fitness (CRF) has been associated with neuroprotective effects among cognitively-intact older adults.

Methods: Twenty-nine aMCI patients underwent MRI for hippocampal structure and function assessment. Associative memory encoding paradigm in which subjects were asked to memorize pairs of face/house-name was used. Two post-scan recognition tasks were performed to evaluate memory performance. CRF (peak oxygen uptake) was evaluated using cardiopulmonary exercise test. Participants were divided into two groups: High-Fit (n = 15) and Low-Fit (n = 14). Between groups analyses were perform on hippocampal volumes, brain activation during memory encoding, and recognition performance.

Results: Although the High-Fit demonstrated greater hippocampal atrophy compared to the Low-Fit (left hippocampus: $p = .002$; right hippocampus: $p = .024$), the groups did not statistically differed in memory performance (recognition 1: $p = .983$; recognition 2: $p = .192$). However, the High-Fit demonstrated greater bilateral hippocampal activation during memory encoding, along with other frontal, parietal, and temporal regions ($p < .05$, FDR corrected). In contrast to these findings, when CRF was controlled for, positive correlations were observed across the whole sample between memory performance and hippocampal structure and function.

Conclusion: CRF in aMCI may be associated with the ability to recruit greater neural resources during memory demands, and by that compensate for increased hippocampal neuropathology in order to preserve cognitive performance.

In: Movement 2019: Brain, Body and Cognition
Editors: Joav Merrick and Gerry Leisman
ISBN: 978-1-53616-714-6

Chapter 93

AUGMENTING LEARNING THROUGH DANCE

Ali Golding[1,*], Dido Green[2] and Ayala Nardi[3]
[1]Department of Dance Science, Trinity Laban, London, United Kingdom,
[2]Department of Rehabilitation, Jönköping University, Jönköping,
Sweden and Royal Free London Hospital, London, United Kingdom
[3]Department of Dance Movement Therapy, Seminar Hakibbutzim, Tel Aviv, Israel

Keywords: dance, motor learning, cognitive function, behaviour

Symposium objective: Professional perspectives and discussion around cognitive-motor interaction in human development and learning through the medium of dance movement, in augmenting cognitive function with a particular focus on complex learning and behaviour challenges.

Symposium summary: Drawing from evidence based models applied in an educational framework {developmental dance movement and autism movement therapy} and clinical practice, professional perspectives will be offered from Ali Golding a dance movement scientist, Ayala Nardi, a dance movement psychotherapist and Dido Green, a pediatric occupational therapist and research therapist with a professional background in dance.

Topics covered include the neuroscience of dance, developmental aspects of cognitive-motor interaction, integration of vestibular and proprioceptive function, rehabilitation of motor dysfunction.

The panel brings innovative research insights from their relevant fields of dance science, dance education and clinical research. Examples include:

Investigating learning through developmental dance movement as a kinesthetic tool in the early years foundation stage.

Expectations and experiences of a dance program for children with autism spectrum disorders: a qualitative study of parents, teachers and therapists.

In conclusion, this symposium proposal brings a much needed platform for the sharing of interdisciplinary approaches between dance in education, occupational therapy and dance movement therapy working to improve complex learning and behaviour challenges potentially impacting upon areas that crossover to the behavioral and communication sciences.

In: Movement 2019: Brain, Body and Cognition
Editors: Joav Merrick and Gerry Leisman
ISBN: 978-1-53616-714-6

Chapter 94

INVESTIGATING LEARNING THROUGH DEVELOPMENTAL DANCE MOVEMENT AS AN EARLY INTERVENTION

***Ali Golding*[1,*] *and Claudia Boes*[2]**
[1]Department of Dance Science, TrinityLaban, London
[2]Department of Occupational Therapy, University of Cardiff, Cardiff, United Kingdom

Keywords: dance, movement, intervention

Aim: Movement is increasingly acknowledged as the primary and dominant means of action, learning and overall development for young children. There is growing evidence that children with learning difficulties frequently present with neurological motor impairment. Presenting background, application and findings pertaining to the relationship between movement and cognition; specifically cognitive-motor interaction in human development through the medium of dance movement with a particular focus on early intervention.

Method: Focusing on impact within an educational framework, Developmental Dance Movement® (DDM) is an example of a model applied in this context. To improve accuracy, validity and give a fuller picture of occurrences of transfer and learning, a mixed method approach was adopted. The investigation is ongoing using a grounded theory approach out of the practice to offer further insights illustrated through case studies.

Results: 2016. Investigating learning through developmental dance movement as a kinesthetic tool in the early years foundation stage. Accelerated learning through observed changed behaviour. T-test results from Aston index pre-post scored drawings showed significant differences ($p = 0.005$) in visual-motor integration and developmental maturity supporting application can provide opportunity for physical/cognitive advancement for young children. In 2015 evaluation of the efficacy of a developmental dance movement program as OT intervention for children with autism spectrum disorder (ASD). Findings indicate the potential for accelerated and cross-curricular learning for all children and suggest that children with SEN and particularly those with autism spectrum disorder (ASD) benefit significantly.

Conclusion: There is evidence of transfer of learning and an increased understanding of students' development needs on the part of the participating teachers. Consequently, the use of DDM as an early intervention to enhance school performance and support other therapeutic goals is promising and an example of interdisciplinary practice between Dance in Education, Occupational Therapy and other professional disciplines extending beyond orthodox health and social care and potentially impacting upon areas that crossover to the behavioral and communication sciences.

In: Movement 2019: Brain, Body and Cognition
Editors: Joav Merrick and Gerry Leisman
ISBN: 978-1-53616-714-6

Chapter 95

EXPECTATIONS AND EXPERIENCES OF A DANCE PROGRAM FOR CHILDREN WITH AUTISM SPECTRUM DISORDERS: A QUALITATIVE STUDY OF PARENTS, TEACHERS AND THERAPISTS

Dido Green[1,*] and Christina Malamateniou[2]
[1]Department of Rehabilitation, Jönköping University, Jönköping, Sweden and Royal Free London Hospital, London
[2]Department of Family Care and Mental Health, University of Greenwich, London, United Kingdom

Keywords: dance, children, autism

Aim: Difficulties with learning and movement are evident for many children with autism spectrum disorders (ASD). The MovementWorks Program, Developmental Dance Movement® (DDM), has been used in schools to promote learning, movement skills and social behaviour. The DDM uses movement therapeutically and as an intentional learning tool; direct instruction of choreographic content provided within meaningful learning contexts. Measurement of effectiveness of specialist movement-dance programs is hampered by lack of sensitivity and responsivity of available tools. This study explores the expectations and experiences of parents, teachers and therapists for children with ASD engaging in a DDM program, in order to develop an evaluation tool aligned with the International Classification of Function core sets for ASD.

* Corresponding Author's Email: dido.green@ju.se.

Method: A qualitative study of a convenience cohort of teachers (n = 6) and parents (n = 2) of children with ASD in the classes receiving DDM and therapists (n = 3) delivering the DDM program. Three role specific focus groups were undertaken to consider potential benefits and challenges of the DDM program for children with ASD. Groups were recorded and transcribed by an independent researcher. Thematic and content analysis was undertaken using NVivo19 to capture manifest and latent meanings.

Results: Three positive themes of behavior, skills and social interaction emerged with differing and overlapping content between therapists, teachers and parents. All groups identified skills of independence, communication and attention/focus, emotional development and physical and sensory development. Teachers and therapists reported engagement, social interaction and cognitive development and therapists and parents reported habits and routines. Therapists, teachers and parents focused differently on stereotypical and restricted behaviours, environmental supports and habits and routines respectively. These themes also emerged as challenges; parents identifying more emotional and behavioural restrictions. A fourth negative theme of transferability of skills and behavior emerged from teachers and therapists.

Conclusion: Using thematic and content analysis 14 items for outcomes were identified by teachers, parents and therapists following a DDT program. These items reflect complex interactions between learning, behaviour and social interaction. Data will be coded against the ICF to develop a specific outcome measure for dance and movement programs for young people with ASD and the tool presented.

In: Movement 2019: Brain, Body and Cognition
Editors: Joav Merrick and Gerry Leisman
ISBN: 978-1-53616-714-6

Chapter 96

MOVEMENT SELF PORTRAITS

Ayala Bresler Nardi*
Therapies Department, Beit Hinoch Aviv, Steiner School, Seminar Hakibbutzim College, Dance Movement Therapy Department Clinical Seminar and Sha'ar Perachia, Tel Aviv, Israel

Keywords: dance, movement, self, learning

Aim: The expression of self is observed through fundamental skills of mobility and walking. Children with learning difficulties often have problems connecting to their bodies which augment deficits in learning within the classroom. The aim of this session will be to explore options for expanding the repertoire of movement expression for children with learning difficulties to enhance learning and behavior.

Method: Multiple case study approach will illustrate techniques to facilitate learning through doing and being with the integration of colours and shapes within rhythmic movement and dance. Qualitative and descriptive approach will be used to describe the process and outcomes of dance movement therapy for children with learning difficulties.

Results: As children were given outlets to move and explore creative movements whilst simultaneously being observed and encouraged in an empathic way, new ways to communicate emerged. Four main themes of self-discovery were identified as children connected with space to describe themselves and their stories. Confidence, self-expression, connectedness and enablement were critical elements emerging through the process to promote not only learning but self-discovery and growth.

Conclusion: Dance movement therapy shows evidence to promote self-concept and communication via facilitative techniques incorporating space and colour within creative movement. These clinical vignettes illustrate the therapeutic principles of grounding to enable children with learning difficulties to become more connected between body and mind with improved confidence and fluency of movement. The impact on social cognition and social behaviour for inclusion with their peers at school will be discussed.

In: Movement 2019: Brain, Body and Cognition
Editors: Joav Merrick and Gerry Leisman
ISBN: 978-1-53616-714-6

Chapter 97

DOCUMENTATION IN MOVEMENT-ORIENTATED MIND-BODY-THERAPIES TAKING THE EXAMPLE OF EURYTHMY-THERAPY

Katharina Gerlach[1,*], Arndt Büssing[1], Erik Baars[2] and Annette Weisskircher[3]

[1]Institute for Integrative Medicine, Faculty of Medicine, University of Witten/Herdecke, Herdecke, Germany
[2]Faculty of Health, University of Applied Sciences Leiden, Leiden, The Netherlands
[3]Institute for Eurythmy Therapy, Faculty of Art-Therapies, Alanus University of Arts and Social Sciences, Alfter, Germany

Keywords: eurythmy-therapy, movement-therapy, mind-body-therapy, documentation, proof of concept, documentation criteria

Objective: As a movement orientated mind-body-therapy, Eurythmy therapy offers therapeutic support for a wide range of symptoms. In anthroposophic medicine it is a key intervention, yet there is little proof of concept. Standardized documentation is a first step towards efficacy evidence. However, documentation of movement-orientated therapies requires appropriate tools, and these tools are only a few. The goal of this project was thus to identify criteria for standardized documentation.

Methods: Literature research in PubMed, Cochrane and Anthromedics databank was made on existing documentation methods for Eurythmy therapy was made. The evaluation of the findings was based on the CARE and CARE-AAT-guidelines, the guidelines on the method of Eurythmy Therapy, the legal basis of the Civil Code, supplementary literature and own reflections.

Results: Strengths and weaknesses of eight documentation methods were described on the basis of formal (coding, patient-perspective, clinical diagnosis), structural (handling and quantitative/qualitative evaluability: survey dates, evidence of effectiveness) and contents (rating of movement qualities, exercises, intentions, development in course of time) criteria. None of them met all criteria as standards in proof of concept.

Conclusion: Standardized documentation must meet certain criteria. This project leads to a working hypothesis on a documentation method that could cover all criteria and is a contribution to the evaluation of movement orientated mind-body-therapies.

In: Movement 2019: Brain, Body and Cognition
Editors: Joav Merrick and Gerry Leisman
ISBN: 978-1-53616-714-6

Chapter 98

THE RELATIONSHIP OF CARDIORESPIRATORY FITNESS AND CREATIVITY IN CHILDREN

Gleydciane Alexandre Fernandes[1], Kell Grandjean da Costa[1], Henrique Bortolotti[1], Isabela Almeida Ramos[2*], Raille Silva de Jesus[1], Karina da Silva Oliveira[3], Tatiana de Cássia Nakano[3] and Eduardo Bodnariuc Fontes[1]
[1]Universidade Federal do Rio Grande do Norte, Natal, Brazil,
[2]Universidade Católica de Brasília, Brasília, Brazil
[3]Pontifícia Universidade Católica de Campinas, Campinas, Brazil

Keywords: creativity, aerobic fitness, cognition

Cardiorespiratory fitness can be improved when children regularly get active playing which to involve aerobic exercises, like tag play, jumping jacks and jump rope. The maximum oxygen consumption (VO2 max) is a validated measurement of cardiorespiratory fitness and usually, children do more aerobic exercises. Lower cardiorespiratory fitness has been widely associated with impaired cognitive performance in children, however, their influence on creativity remains unclear. Creativity is a component of cognition and is defined as the process of identifying the difficulty, formulating hypotheses about the deficiencies and solving the problems. For instance, creativity is suggested to contribute for personal and professional success that can be important for children's development.

Objective: Here, we compared creativity in children with different levels of cardiorespiratory fitness (CF) and their relationship.

Methods: In this cross-sectional study, 73 children with age ranging from 10.1 to 11.5 years performed two testing sessions. On the first testing session, all children had general anthropometric assessments and completed a shuttle run test to estimate cardiorespiratory fitness (VO2 max). On the second session, they perform the Children's Figural Creativity Test (CFCT), scholar performance test and have the fat mass (FM) measured by the Dual-energy X-ray Absorptiometry. Children were divided into two groups (Lower x Higher) separated by the median value of CF. Creativity was compared between groups by a paired t-test and unpaired Mann-Whitney ($P < 0.005$). Pearson or Spearman correlations were used to compare the associations between the variables.

Results: The results demonstrate no significant differences in creativity between groups with higher and lower VO2 max ($p = 0.75$). In addition, no correlation was identified between creativity and CF ($r = -.031$; $p = .828$).

Conclusion: In our study, the CFCT showed no difference in creativity between groups with different VO2 levels. No relationship was found between variables of the Children's Figural Creativity Test variables with CF. These could be related to low sample size, indirect measure of CF by the shuttle run, or lack of neurobiological markers. Therefore, more studies are necessary with larger sample sizes, a laboratory-based test of VO2 max, or neurobiological markers (NIRS, fMRI).

In: Movement 2019: Brain, Body and Cognition
Editors: Joav Merrick and Gerry Leisman ISBN: 978-1-53616-714-6

Chapter 99

SELF-GENERATED AUDITORY CUES ENHANCE GAIT KINEMATICS FOR PEOPLE WITH PARKINSON DISEASE

Elinor C Harrison[1,*], Adam P Horin[1], Marie E McNeely[1-3], Peter S Myers[1], Kerri S Rawson[1], Ellen N Sutter[1,4] and Gammon M Earhart[1,2,5]

[1]Program in Physical Therapy, Washington University School of Medicine, St. Louis, Missouri, United States of America
[2]Department of Neurology, Washington University School of Medicine, St. Louis, Missouri, United States of America
[3]Unfold Productions, LLC, St. Louis, Missouri,
[4]Department of Rehabilitation Medicine, University of Minnesota, Minneapolis, Minnesota, United States of America
[5]Department of Neuroscience, Washington University School of Medicine, St. Louis, Missouri, United States of America

Keywords: rhythmic auditory cueing, self-generated, singing, Parkinson disease, gait kinematics

Objective: Self-generated cueing techniques, e.g., singing or mental singing, may provide greater benefit to gait for people with Parkinson disease (PD) as compared to external cueing techniques, e.g., listening to music, by eliminating the need to match an external source. Mental singing, in particular, can improve spatiotemporal features such as gait variability, but the effects on gait kinematics are unknown. In this study, we

* Corresponding Author's Email: harrisonel@wustl.edu.

sought to compare the effects of different rhythmic cues on lower limb movement trajectories during gait in people with PD.

Methods: 35 participants with idiopathic PD (21 male) were recruited and tested 'ON' medication. Sagittal plane joint angles of the hip, knee, and ankle were collected for 3 trials in each of 4 gait conditions (UNCUED, MUSIC, SING, MENTAL) at 110% of preferred cadence. Song tempo was adjusted for each participant. Repeated measures MANOVAs were used to assess differences between gait characteristics, gait variabilities, and joint ranges of motion. Pearson Chi-Square assessed differences between responders (i.e., those who increased gait velocity in MENTAL by > .06m/s) and non-responders. Post-hoc pairwise comparisons were used as appropriate. Statistical significance was set at a < .05.

Results: A main effect of condition on spatiotemporal gait characteristics (F (9, 306) = 7.328, p < .001) revealed that velocity (p < .001), cadence (p < .001), and stride length (p = .029) were higher in cued conditions than in UNCUED. A main effect of condition on gait variability revealed that stride time variability (p = .003) was lower in MENTAL than MUSIC (p = .005) and SING (p = .013). A main effect on total hip ROM (F (3, 102) = 11.647, p < .001) showed increased ROM in cued conditions compared to UNCUED (all p < .003). Responders (n = 23) improved velocity, stride length, and stride length variability (all p < .01) more than non-responders.

Conclusion: This study provides important insight into how self-generated cueing techniques may benefit gait stability in PD, potentially reducing the risk of falls. Mental singing holds promise as an effective alternative to external cueing for some people with PD. More work is warranted to determine what factors contribute to likelihood of responding positively to self-generated cues.

In: Movement 2019: Brain, Body and Cognition
Editors: Joav Merrick and Gerry Leisman
ISBN: 978-1-53616-714-6

Chapter 100

SING WHILE YOU STRIDE: A WORKSHOP COMBINING MOVEMENT WITH SONG

Elinor C Harrison*

Program in Physical Therapy, Washington University School of Medicine, St. Louis, Missouri, United States of America

Keywords: dancing, singing, motor control, movement disorders

Objective: Moving and singing are inextricably linked, and recent evidence suggests that doing them simultaneously can improve the performance of both. In this workshop, participants will learn some of the neuroscientific evidence showing the linkage between these two seemingly disparate skills. By applying principles of vocal-motor coupling, we will explore how vocalizing while you move may provide a powerful tool to aid movement in healthy adults and people with neurodegenerative disorders. In combining vocal and somatic exercises, we will learn techniques utilized by both singers and dancers and discover new tools to help us all find our voices.

In: Movement 2019: Brain, Body and Cognition
Editors: Joav Merrick and Gerry Leisman
ISBN: 978-1-53616-714-6

Chapter 101

THE SIGNIFICANCE OF THE VESTIBULAR FUNCTION IN CHILD DEVELOPMENT

Tatjana Adamovic**, Mirjana Sovilj and Ljiljana Jelicic
Institute for Experimental Phonetics and Speech Pathology,
Belgrade and Life Activities Advancement Center, Belgrade, Serbia

Keywords: vestibular sense, newborn, balance ability, child, development

Objective: Longitudinal study was carried out with dual goals: to establish whether the success rate in performing reflexes related to the function of the vestibular sense at birth is in correlation with the success rate in maintaining balance in early childhood; and to establish whether the vestibular function can be observed as a new maturation parameter.

Methods: The research sample comprised N = 54 children of both genders, aged 5.0 to 5.4, whose balance ability (BA), speech and language development (SLD), sensory-motor development (SMD) and social-emotional behaviour (SEB) were evaluated, based on the battery of standardized tests. The group of reflexes, related to the function of the vestibular sense, were previously clinically tested on the third day upon birth, in all children from the sample. The data at birth and at the age of 5 were recorded by a digital camera, subsequently scored and statistically and descriptively processed. In addition, data on the time when children included in research start to sit, stand, walk and talk (SSWT), were taken for each examinee individually.

Results: The research results indicated a statistically significant correlation between the presence of all four reflexes collectively at birth (Moro reflex + Head righting reflex

* Corresponding Author's Email: tadus3@gmail.com.

+ Asymmetrical tonic neck reflex + Vestibulo ocular reflex), and the results of the development of the BA (r = + 0,50, level 0,01), SLD (r = + 0,51, level 0,01), SMD (r = + 0,44, level 0,01) and SEB (r = + 0,38, level 0,01) at the age of 5. It was also established that children with better BA at the age of 5, begin with SSWT significantly earlier.

Conclusion: The significance of this research is reflected in the new knowledge in the domain of maturation and function of the sense of balance immediately upon birth, since this segment of a newborn's physiology has never been processed in a similar way. Balance ability can be influenced by targeted exercises, and this study indicated the way to identify which children are in need of such exercises.

Acknowledgment: Research supported by the Ministry of Education, Science and Technological development of the Republic of Serbia within the project no. 178027.

In: Movement 2019: Brain, Body and Cognition
Editors: Joav Merrick and Gerry Leisman
ISBN: 978-1-53616-714-6

Chapter 102

ASSOCIATIONS BETWEEN BACKACHE AND PERSONAL VULNERABILITY AMONG ISRAELI SCHOOLCHILDREN

Liat Korn, Noa Ben Ari, Riki Tesler* and Yosi Harel Fisch
Ariel University, Ariel and Bar Ilan University, Ramat Gan, Tel Aviv, Israel

Keywords: backache, school children, social support, academic stress, parental communication

Objectives: Backache is the leading cause of disability worldwide, reported among all age groups from children to the elderly. Our objective was to investigate associated factors for backache among children in Israel

Methods: An HBSC structured, anonymous, self-report questionnaire was distributed in 2014. The representative sample included 3594 participants, 49.2% of them female, from 6th, 8th, and 10th grades, studying in Israel's three education sectors: Jewish secular (39.2%), Jewish religious (35.1%), and Arab (25.7%). The questionnaire included validated questions on various subjects related to health status and health behaviours.

Results: The risk of backache was higher when participants reported smoking (AOR = 2.25; $p < 0.001$) or experiencing less parental communication (AOR = 1.68; $p < 0.001$), less social support (AOR = 1.31; $p < 0.05$), or academic stress (AOR = 1.27; $p < 0.05$).

Conclusion: Smoking, lack of parental communication, less social support, and academic stress are significantly stronger measures associated with backache among children and adolescents than BMI, physical activity, or sedentary behavior.

In: Movement 2019: Brain, Body and Cognition
Editors: Joav Merrick and Gerry Leisman
ISBN: 978-1-53616-714-6

Chapter 103

ASSOCIATIONS BETWEEN SELF-REPORT PHYSICAL ACTIVITY AND PERCEIVED SCHOOL PERFORMANCE OF YOUNG ADOLESCENTS IN ISRAEL

Kwok Ng[1,2], Riki Tesler[3,*] and Yossi Harel-Fisch[4]
[1]School of Educational Sciences and Psychology,
University of Eastern Finland, Joensuu, Finland
[2]Department of Physical Education and Sport Sciences,
University of Limerick, Limerick, Ireland
[3]Department of Health Systems Management, Ariel University, Ariel, Israel
[4]School of Education, Bar-Ilan University, Ramat Gan, Israel

Keywords: transition, academic performance, sport, adolescence

Objectives: Regular physical activity and doing well in school are important for growing adolescents. The evidence in physical activity for health for adolescents suggests daily moderate-to-vigorous physical activities (MVPA) of at least 60 minutes. There is much debate about how MVPA is associated with academic performance, particularly in the school setting. In this study, the complexities of adolescents' development and their relationship to physical activity and school performance are examined together across young adolescents in Israel.

Methods: Data on young adolescents from Israel (n = 5,121) were investigated in the associations between MVPA and the perception of school performance (PSP). One single item on MVPA and another on PSP were used from the WHO collaborative cross-national Health Behaviour in School-aged Children (HBSC) study. There were four

* Corresponding Author's Email: kwok.ng@hbsc.org.

MVPA groups for multi-nominal analyses; none (0-2 days) – reference group, low (3-4 days), moderate (5-6 days), active (7 days). There were also four PSP groups; below average (reference); average, good, and very good. Stratification and analyses by sex were carried out to indicate the changes of age cohorts of 11y, 13y, and 15y olds.

Results: Although boys were more physically active than girls ($p < .001$), more girls reported very good PSP than boys ($p < .001$). For both boys and girls, the highest odds for very good PSP were among adolescents who reported 5-6 days MVPA, and there were no significant associations among adolescents with average or good PSP with daily (7 days) MVPA when compared with below average PSP and 0 days MVPA.

Conclusion: There were differences between boys and girls with MVPA and PSP. Moreover, there are effects of age on adolescents' PSP and MVPA, although together patterns remain relatively similar. Appropriate health promotion approaches may need to consider the child's current levels of school performance and level of physical activity.

In: Movement 2019: Brain, Body and Cognition
Editors: Joav Merrick and Gerry Leisman ISBN: 978-1-53616-714-6

Chapter 104

PERCEPTION OF TEACHERS OF BASIC EDUCATION ABOUT MOVEMENT, COGNITIVE ACTIVITIES AND NEEDS OF CHILDREN DEVELOPMENT

Rute Estanislava Tolocka**, Wagner Porto Rego and Vitor Antonio Cerignoni Coelho

Department of Health Sciences, Methodist University, UNIMEP, Piracicaba, São Paulo, Federal Institute of Education and Sciences and Faculty of Physical Education, IFBA, Bom Jesus da Lapa, Bahia and Federal University of Tocantins – Miracema, Tocantins, Brasil

Keywords: movement, cognition, children development, school

Objective: The aim of this study was to verify teacher's valorisation to movement activities (MA) and cognitive activities (CA) in order to improve child development and frequencies of such activities.

Methods: 70 teachers answered a questionnaire about importance, frequency and needs of MA and CA for children development. The validation was done with Crombach's Alpha; associations were checked through Kruskal Wallis test and multiple linear regression analysis.

Results: MA and CA were considered essential for children by 40% and 36% of the sample but daily frequency of MA and CA were reported by 12% and 30%; the necessity of MA and CA for children development were referred by 5,7% and 15,72% respectively. Multiple linear regression analysis adjusted for potential confounders (i.e., age,

* Corresponding Author's Email: tkiva05@yahoo.com.br.

socioeconomic status) revealed relationship between importance and necessity of MA, with X = 90, 2 for $p < 0, 01$ and frequency of CA and necessity of MA. Kruskal Wallis test confirmed the correlation between importance and necessity of MA.

Conclusion: There is association between necessity of MA for children development and frequency of MA offered for children, both with low scores, i.e., teacher offer few opportunities to movement activities and did not think it is important to the development. Although teachers think that both MA and CA are important for children, they did not recognize that MA is linked with CA or with necessity of children development. It is necessary to spread the link between movement and cognition that has been demonstrated in scientific studies to improve opportunities for children experiment more MA and CA in order to reach optimal development.

In: Movement 2019: Brain, Body and Cognition
Editors: Joav Merrick and Gerry Leisman
ISBN: 978-1-53616-714-6

Chapter 105

EFFECT OF ACUTE ACTIVE PLAYING ON NEUROELECTRIC MEASURES IN CHILDREN

Isabela Almeida Ramos**, *Stéphany Vieira Brito, Raiane Maiara dos Santos Pereira, Gustavo Lira Guedes, Eduardo Bodnariuc Fontes and Carmen Silvia Grubert Campbell
Catholic University of Brasilia, University Center Euro American – UNIEURO and Federal University do Rio Grande do Norte, Natal, Brazil

Keywords: ERPs, children, active play

Event-related brain potentials (ERPs) have been instrumental for discerning the relationship between physical activity and aspects of cognition in children.

Objective: to verify and compare the waveform of N450 ERP component in Stroop task between active play, passive play and control sessions in children.

Methods: 13 girls and boys (9.5 ± 0.8yr; 32.05 ± 5.9kg; 1.38 ± 0.1m; 16.36 ± 4.2%body fat) participated on this study. Firstly, they were performed anthropometric evaluations. Thereafter, three sessions lasting 30min were performed in randomized order: 1) Control – seated watching a documentary (92.4 ± 8.9bpm); 2) Active Play – games with movement, e.g. jumping jacks, running, jumping (152.8 ± 16.31bpm); 3) Passive Play – playing games seated, e.g. puzzles of colors, animals and numbers (99.6 ± 28.2bpm). Subsequently, they completed a Stroop task (GO/No-go) while brain electrical activity was recorded with EEG. ERPs were recorded and calculated from electrodes at

* Corresponding Author's Email: ahbeuramos@gmail.com

Fz, F3, F4, Cz, C3, C4, Pz, P3 and P4. Mean peak amplitude of N450 component were used to compare the electrical brain responses between sessions.

Results: The ANOVA of N450 mean amplitudes revealed a significant main effect for interaction of factor frontal anterior between sessions $F(2,36) = 3.63$, $p = 0.04$ (fig 1). The N450 amplitude in the incongruent condition was less negative in active play than in control session in the frontal sites ($p = 0.11$). In the Incongruent condition, the mean of Reaction Time was slower to the Active Play (859ms) than for Control (886ms) and Passive Play (893ms), ($p = 0.84$).

Conclusion: 30 min of active play could contributed to improve cognitive performance, in the component N450 about incongruent condition, in children.

Supported by CNPq and CAPES.

In: Movement 2019: Brain, Body and Cognition
Editors: Joav Merrick and Gerry Leisman
ISBN: 978-1-53616-714-6

Chapter 106

LEARNING NEUROSCIENCE THROUGH MUSIC, DANCE AND ACTIVE PLAYING

Isabela Almeida Ramos*[*], *Raiane Maiara dos Santos Pereira*, *Gleydciane Alexandre Fernandes*, *Gustavo Lira Guedes* and *Carmen Silvia Grubert Campbell
Catholic University of Brasilia, University Center Euro American–UNIEURO
and Federal University do Rio Grande do Norte, Natal, Brazil

Keywords: neuroscience, dance, aerobic exercise

It is well-known that regular physical activity ameliorates several cognitive abilities in school children. Through movements, dancing and music is possible improves brain health, executive functions and learning. Previously we compared the effects of a physical education program combined with scholar contents on selective attention in boys and girls, and it has shown positive effects. The aim of this lecture is to demonstrate a different way to learn neuroscience through dance, music and physical exercise, and improves cognitive functions. Firstly, the teacher will perform a music wrote from her doctoral dissertation, about the effects of acute and chronic physical exercise on event-related brain potentials in children. Then, two groups of 10 participants will be created, each group will introduce themselves, followed by stretching and warm up, they will participate on active plays involving aerobic exercises including front squats, push-ups, jumping jacks and music with neurocognitive lyrics like a parody, and others activities with rhythm. Lastly, all of them will participate on treasure hunt of physical fitness, to find the treasure is necessary solve puzzles with some aerobic exercises and answer

[*] Corresponding Author's Email: ahbeuramos@gmail.com.

questions about what they have learned at workshop, then, they will run away looking for the box. The first place team will win a brain treasure's box. We hope to shown different and attractive strategies to play and learn through this intervention.

In: Movement 2019: Brain, Body and Cognition
Editors: Joav Merrick and Gerry Leisman
ISBN: 978-1-53616-714-6

Chapter 107

THE APPLICATION OF THE SYNCYTIOTROPHOBLAST FOR THE TREATMENT OF ATHEROSCLEROTIC PLAQUE

Enzo Santiago Rezende**, *Ana Lídia Ferreira de Oliveira and Clara Carolina Souza Santos
Department of Medicine, Faculdade de Tecnologia e Ciências, Salvador, Bahia and Department of Medicine, Faculdades Santo Agostinho, Vitória da Conquista, Bahia, Brazil

Keywords: syncytiotrophoblast, atherosclerotic, treatment

Objective: Currently, there is a large number of people with a sedentary lifestyle and inadequate nutrition. Alongside with these, the lack of physical activity, obesity, smoking, hypertension, and diabetes have been increasing the incidence of atherosclerosis. This condition is generated due to the accumulation of lipids in the vessels, which form an atheroma (plaques composed specially by lipids and fibrous tissue that are formed on vessels' walls), compromising the blood circulation and thereafter causing cardiovascular diseases, responsible for 30% of deaths in Brazil.

Method: Bibliographic analyzes informs us that during the implantation of the embryo in the endometrium of the woman, a type of cell called syncytiotrophoblasts digests decidual cells, that are rich in glycogen and lipids, thence creating room for the implantation of the embryo. Having such knowledge, as atheroma's are lipid-rich, it is intended to apply the syncytiotrophoblasts to blood vessels that were harmed by this

* Corresponding Author's Email: enzo-santiago@hotmail.com.

condition, in order to cause a disruption in such vessels and create a way to make the activity cease shortly after the digestion of the of the plaque, making the blood flow normally in that vessel

Results: Nowadays, the treatment for people suffering of atherosclerosis is, in most cases, a surgical procedure, an invasive process. About one million people are hospitalized with cardiovascular complications each year in Brazil, increasing the demand for surgeries, and leaving many people in the waiting rows. The decrease of the number of people waiting for surgical procedures is important for the whole public health system, as it would cut expenses, having the same efficiency to the treatment.

Conclusion: Due to a larger number of people being exposed to several risk factors, the number of cases of atherosclerosis has been increasing over the country. Since its most common treatment is the surgical intervention, it is mandatory to look for a faster and equally efficient treatment. In pursuit of this purpose, the application of the syncytiotrophoblasts into the vessels affected by atheroma can be an effective and safe treatment for those patients who can suffer more serious diseases, and even death.

In: Movement 2019: Brain, Body and Cognition
Editors: Joav Merrick and Gerry Leisman

ISBN: 978-1-53616-714-6

Chapter 108

PREFRONTAL CORTEX FUNCTIONING IN OLDER ADULTS: PRELIMINARY NEUROBEHAVIORAL FINDINGS INDICATING PRODROMAL ACCELERATED COGNITIVE DECLINE IN AGING INDIVIDUALS

***Jonathan David**, *Braw Yoram and Oded Meiron*†**
Clinical Research Center for Brain Sciences, Herzog Medical Center, Jerusalem
and The Department of Behavioral Sciences, Ariel University, Ariel, Israel

Keywords: ageing, executive attention, auditory event-related potentials (ERP's), Working memory, dementia

Objective: The goal of the current study was to explore associations between cognitive decline and aging in healthy individuals and to possibly detect a subgroup that indicates extremely reduced verbal and echoic working memory functioning within the older sample (ages-45-90). In accordance, we aimed to identify cases that attained poor scores on both executive attention (EA) functioning, and EEG-based "pathological aging" biomarker. The EEG biomarker is extracted from mismatch negativity (MMN) auditory event-related potentials (ERPs). Both EA and MMN represent neurobehavioral measures of prefrontal cortex (PFC) excitability, and MMN is directly related to N-methyl-D-aspartate (NMDA) receptor function. Importantly, we intended to identify "high-risk" (HR) criterion group, representing PFC-mediated accelerated neuro-

* Corresponding Author's Email: Jonathandavid9@gmail.com.
† Corresponding Author's Email: oded@herzoghospital.org.

physiological and cognitive decline, hypothesized to precede dementia, and other age-related neurological disorders (e.g., Parkinson's disease).

Methods: Thirty-three participants 18-90 years old were examined. WM accuracy and MMN ERP components were assessed across all participants. HR individuals were defined as participants that their scores were in the range of the lowest interquartile (IQR) on both MMN amplitude and WM accuracy score-distributions. Later, to investigate possible "prefrontal-functioning" markers for classifying clinically HR individuals, a comparison was made between groups of young (age18-45), older (age 46-90), and "HR" sub-group participants (age 46-90).

Results: Findings demonstrated a significant correlation between age and WM accuracy on the EA task ($r = -.46$, $p < .001$). Furthermore, a significant correlation between age and MMN amplitudes (i.e. less amplitude negativity as we age) was found ($r = .63$, $p < .001$). Six older subjects were putatively defined as HR group displaying 1st IQR scores on both MMN and EA measures. Only WM-RT's on the EA task significantly differentiated older-HR participants versus normal older, and versus young participants ($F (2, 27) = 14.20$, $p < 0.01$).

Conclusion: Current findings promote the sensitivity of functional PFC measures to detect age-related cognitive decline in aging adults. More so, our findings, after longitudinal validation, could characterize clinically HR individuals susceptible to late-life neurodevelopmental disorders associated with widespread dysfunctions in the NMDA receptor system.

In: Movement 2019: Brain, Body and Cognition
Editors: Joav Merrick and Gerry Leisman
ISBN: 978-1-53616-714-6

Chapter 109

ADAPTED BALLROOM AND LATIN DANCE PROGRAM: A PRIMER FOR APPLICATION IN PEOPLE WITH MOVEMENT DISORDERS IN ORDER TO IMPROVE BALANCE, FUNCTIONAL MOBILITY, COGNITION AND THE QUALITY OF LIFE

Rita Bindar Elimeleh*[*] *and Yulia Gamerman
"Dance through life" project, Ramat Gan, Rishon Letzion, Ramat Hasharon, Israel

Keywords: dance, rehabilitation, elderly, PD, therapy, movement, music

Objective: Movement strategies, like dance, requiring strong cognitive involvement and intentional motor planning have been demonstrated to result in improved movement quality in people with Parkinson's disease. Recently, cognitive benefit, particularly in visuospatial domains have also been documented. The purpose of "Dance through life" program is to help those elderly individuals with balance impairments, movement disorders and PD patients to slow the progression of the disease, improve balance, functional mobility, cognition and quality of life

Methods: the combination of Physiotherapy with ballroom and Latin dance is used to meet the therapeutic goal of the project. There are structured group lessons instructed by dance movement therapist with large experience in dance instruction and physical therapist. For rehabilitative purposes of the people with postural impairments some aspects of partnered dances need to be adapted for the best effect. This experiential

[*] Corresponding Author's Email: ritabindar@yahoo.com.

workshop will present the principles of adapted ballroom and Latin dances, such as cha cha cha and foxtrot, and explore the extrapolation of these principles to various neurological conditions, e.g., Parkinson's disease (PD), multiple sclerosis (MS). We will explain how to adapt dances for different ability levels and with respect to the symptoms of different disease, suggesting reasonable treatment goals that can be achieved with partnered dance methods. We will demonstrate how to build dances from everyday actions to explore imagination, expression and improvisation.

This particular program has proved itself to have a great impact on the lives of people with PD, MS and other neurological and movement impairments. It has unquestionable benefits for the participants and may prolong the life spans and improve the quality of life.

In: Movement 2019: Brain, Body and Cognition
Editors: Joav Merrick and Gerry Leisman
ISBN: 978-1-53616-714-6

Chapter 110

THE UTILITY OF CLINICAL PILATES METHOD AS A TOOL FOR CLINICIANS WITH THE PURPOSE OF MENTAL AND PHYSICAL HEALTH, TRAUMA PREVENTION AND REHABILITATION

Eleni Matsouki*
First page affiliation

Keywords: clinical pilates, spine mobility, body awareness, rehabilitation

In this workshop, the participants will learn how to apply the six principals of Clinical Pilates Method in every kind of exercise they use in their programs, regardless if they aim to empowerment, prevention or rehabilitation. The first principle is concentration, which promotes the mindful movement of the body. The next two principles are the pilates breathing and the activated core (or 'powerhouse', according to the creator of the method himself, Joseph Pilates), which are always combined and together succeed strengthening the deep core muscles, lengthening the spine, maintaining the right posture and maximizing flexibility of the body. The last three principles refer to the quality of the movement. These are the motion control, precision and flow.

Moreover, the participants of this workshop will be taught how to use specific exercises from the pilates repertoire in the assessment of joint range of motion and the muscle strength. Through this procedure, which is a kind of mechanical examination, the participants will be able to identify restrictions in the movement of the spine and the peripheral joints, as well as soreness or restraints in the muscles. If the above assessment is combined with the scientific knowledge the participants already possess, they will be able to come to save conclusions as to their patients' psychological and somatic status, and decide what kind of exercises will include in their patients' program. All exercises can change level of difficulty by the use of small equipment that can be found in every clinic, such as stability balls, bands, rings, pillows, etc. The advantage of this method is that it can be performed by people of all ages, no matter what their physical or cognitive condition is.

Conclusively, the Clinical Pilates Method constitutes important knowledge for all those who deal with the mental health and the movement of the human body.

In: Movement 2019: Brain, Body and Cognition
Editors: Joav Merrick and Gerry Leisman
ISBN: 978-1-53616-714-6

Chapter 111

NOVICE WALKERS SOLVE A MOTOR CHALLENGE: DOES SLEEP QUALITY PREDICT PERFORMANCE?

***Tamar Simon*[1,*], *Melissa N Horger*[2], *Sarah E Berger*[2] *and Anat Scher*[1]**
[1]Department of Counseling and Human development, University of Haifa, Haifa, Israel
[2]Department of Psychology, The College of Staten Island, CUNY,
Staten Island, United States of America

Keywords: infants, motor learning, sleep, nightwaking

In adults, naps and nighttime sleep impacts motor learning. The contribution of naps to infants' motor learning has been demonstrated, however, if and how nocturnal sleep quality affects infants' preparedness for next day learning has not been addressed.

Objective: The current study examines the role of sleep on motor problem solving in infants who just started to walk independently. This stage of development provides a unique window to examine coping with the motor challenge of shifting from up-right posture into crawling through a tunnel - a demanding task for novice walkers. Drawing on evidence that young children with consolidated and well-regulated sleep perform better on cognitive tasks, we hypothesized that good sleep will positively contribute to solving a motor challenge encountered by novice walkers.

Methods: The tunnel test (2-meter-long lightweight colorful toy tunnel) was administered to 30 infants, within one week of giving-up crawling. On the night preceding the test, infants' sleep was recorded using actigraph - user friendly wrist-watch like device that provides a measure of sleep quality. An algorithm validated for infants,

* Corresponding Author's Email: tamarsim@gmail.com.

provides sleep measures, including, duration, nightwaking, and overall sleep quality. Tunnel performance was filmed and scored for number of steps to complete the task - maximum of 15 prompts using a strict protocol; duration; exploratory postures - shifts in body position before entering the tunnel; number of postural shifts while in the tunnel. Inter-rater agreement has been established (>.90%).

Results: Preliminary findings document a link between sleep fragmentation (e.g., nightwaking) and performance. More nightwaking episodes were associated with longer latency at the tunnel entrance and more postural shifts. The results indicate that sleep quality, rather than duration, is associated with infants' daytime readiness to learn.

Conclusion: This is the first study to demonstrate that sleep quality contributes to infants' motor learning and expands the limited empirical evidence on the importance of good night sleep for efficient learning as infant's transition to toddlerhood.

This work was supported by National Science Foundation grant BCS-1551703 to SEB, and Binational Science Foundation (BSF-2015606) to AS.

In: Movement 2019: Brain, Body and Cognition
Editors: Joav Merrick and Gerry Leisman
ISBN: 978-1-53616-714-6

Chapter 112

OXYGEN: A LIMITING FACTOR FOR BRAIN RECOVERY AND ENHANCED COGNITIVE-MOTOR PERFORMANCE

Shai Efrati*
Sagol Center for Hyperbaric Medicine and Research, Assaf Harofeh Medical Center, Zerifin, Sackler School of Medicine and Sagol School of Neuroscience, Tel-Aviv University, Tel-Aviv, Israel

Keywords: oxygen, brain metabolism, cerebral blood flow

Effective brain metabolism is highly dependent on a narrow therapeutic window of oxygen. There is a linear relationship between normal arterial pO_2 and brain tissue oxygenation ($PbTO_2$). At normal, healthy conditions, brain metabolism reaches the upper limit of oxygen consumption, which makes it dependent on cerebral blood flow (CBF). At each time point the CBF shifts to more active regions (task dependent) at the expense of other, less active regions. Those, oxygen delivery is a rate limiting factor for brain activity at normal healthy individuals and more so in chronic non recovering brain injury.

Clinical studies published this year present convincing evidences that hyperbaric oxygen therapy (HBOT) can be the coveted neurotherapeutic method for brain repair. The improved brain tissue oxygenation and oxygen fluctuations generated during HBOT induced neuroplasticity mediated by HIF, VEGF, stem cells stimulation and brain angiogenesis and neurogenesis. The clinical beneficial effects are motor and/or cognitive, depending on the recovered part of the brain.

The understanding that oxygen is a limiting factor even for the healthy brain activity, drives the motivation to use HBOT for enhancing the "normal" performance. The results of the new studies, where HBOT is used to enhance cognitive and physical performance of "elite" sportsman, are very intriguing.

In this lecture we will discuss the new understanding of the multi-faceted neurotherapeutics role of HBOT in both health and disease conditions.

In: Movement 2019: Brain, Body and Cognition
Editors: Joav Merrick and Gerry Leisman
ISBN: 978-1-53616-714-6

Chapter 113

THE RELATIONSHIP BETWEEN VOLUMETRIC MRI AND MOVEMENT VARIATION AND ADAPTABILITY IN PRETERM INFANTS

Ana Katušić[1,2,*], Marina Raguž[3], Iris Žunić Išasegi[1] and Ivica Kostović[1]

[1]Croatian Institute for Brain Research, Centre of Excellence for Basic, Clinical and Translational Neuroscience, University of Zagreb, School of Medicine, Croatia
[2]Day-care Centre for Rehabilitation Mali dom-Zagreb, Zagreb, Croatia
[3]University Hospital Dubrava, Department of Neurosurgery, University of Zagreb, School of Medicine, Croatia

Keywords: brain volumetry, motor development, movement variation, adaptability, infant motor profile, preterm infants

Objective: Premature birth has been associated with poor neurodevelopmental outcomes, including motor functions. Lack of movement variations occur even in the absence of significant structural brain lesions and have been related to more subtle alterations in brain development. In order to analyze relationship between brain structure and motor functions, volumetric measurements performed on magnetic resonance imaging (MRI) in a cohort of preterm infants was correlated with movement variation and adaptability at 12 months corrected age (CA).

Methods: Forty preterm infants born between 24 and 32 weeks gestational age (GA) underwent MRI with volumetric analysis at term-equivalent age (TEA). MNI Toolbox program was used for the analysis of high resolution magnetization-prepared rapid acquisition gradient echo (MPRAGE) images and tissue segmentation based on the intensity differences in-between the voxels. Infants were assessed using the Infant Motor Profile (IMP) at 12 months CA in order to evaluate variation and adaptability of movements and motor performance.

Results: The volume of frontal lobes, deep gray matter (basal ganglia and thalamus) and cerebellum were significantly reduced in the children with lower scores in variation and adaptability of movements. Smaller deep gray matter ($p < .02$) and cerebellum ($p < .04$) volume were significantly associated with poorer motor performance at 12 months CA. No differences were detected in other volumetric measurements of brain structures.

Conclusion: There is an association between diminished volume of the frontal lobe, deep gray matter and cerebellum volume and reduced movement variation and adaptability. Our data suggest that volumetric MRI preformed at TEA can provide an additional structural parameter for early prediction of motor development in preterm infants.

In: Movement 2019: Brain, Body and Cognition
Editors: Joav Merrick and Gerry Leisman
ISBN: 978-1-53616-714-6

Chapter 114

EFFECTS OF TANGO ARGENTINE ON THE RELATIONSHIP EXPERIENCE IN PATIENTS WITH PARKINSON'S DISEASE (PD) FROM THE PERSPECTIVE OF THE PATIENT

Yvonne Beerenbrock[1], Sunnyi Mews[2], Lea Meyer[2], Stephanie Herrlich[2], Juliane Böhme[2], Bettina Berger[3], David Martin[3] and Arndt Büssing[1,*]

[1]Quality of Life, Spirituality and Coping, Institute for Integrative Medicine, Faculty of Health, Witten/Herdecke University, Herdecke, Germany
[2]Department of Psychology and Psychotherapy, Faculty of Health, Witten/Herdecke University, Herdecke, Germany
[3]Medical Theory, Integrative and Anthroposophic Medicine, Faculty of Health, Witten/Herdecke University, Herdecke, Germany

Keywords: Parkinson's disease (PD), Tango Argentine, relationship experience, patient's perspective

Objective: As a chronic disease, PD influences all areas of life of those affected. In addition to motor symptoms, there are also impairments in the economic, emotional and social spheres of life. The dancing of the Argentine tango seems to have a special effect on the couple relationship as a holistically appealing form of movement-supported intervention. However, there are hardly any publications on this subject.

Question: Does the Tango Argentine have a perceptible effect on the relationship from the perspective of the patient?

Method: Qualitative interviews (based on semi-structured guidelines) of 11 affected patients who were participating Tango Argentine courses for persons with PD. The interview material was examined according to content analysis criteria (according to Mayring) and thematically coded (according to Flick).

Results: From the interviews the following seven main categories have been generated: 1. relationship in general (re-experience and further development of emotional attachment), 2. understanding/support and acceptance from the partner (verbal/non-verbal communication), 3. own or common body experience (realistic correction of the symptoms), 4. dealing with each other (appreciation), 5. common experience (special kind of communication), 6. tango beyond the course (social exchange), 7. distribution of roles and leadership in partnership (positive effect on self-efficacy).

Conclusion: Tango Argentine has a noticeable influence on the psycho-emotional experience of PD patients. An essential dimension of the tango is the non-verbal communication about body tension, change of direction and eye contact as well as the guidance through the affected person, which leads to a re-experience of emotional attachment. Therapeutically relevant, the mindfulness-based aspects of dance lead both partners from the disease-limiting relationship experience into a resource-oriented perception of the relationship.

In: Movement 2019: Brain, Body and Cognition
Editors: Joav Merrick and Gerry Leisman
ISBN: 978-1-53616-714-6

Chapter 115

The effects of Tango Argentine on body experience in patients with Parkinson's disease (PD)

Yvonne Beerenbrock[1], Lea Meyer[2], Stephanie Herrlich[2], Juliane Böhme[2], Sunnyi Mews[2], Bettina Berger[3], David Martin[3] and Arndt Büssing[1,*]

[1]Quality of Life, Spirituality and Coping, Institute for Integrative Medicine, Faculty of Health, Witten/Herdecke University, Herdecke, Germany
[2]Department of Psychology and Psychotherapy, Faculty of Health Witten/Herdecke University, Herdecke, Germany
[3]Medical Theory, Integrative and Anthroposophic Medicine, Faculty of Health, Witten/Herdecke University, Herdecke, Germany

Keywords: Parkinson's disease (PD), Tango Argentine, body experience

Objective: Patients with PD experience somatic and psycho emotional limitations due to illness. As a psychosomatic factor, subjective body experience clearly determines disease experience. Patients with PD have a significantly more negative body experience than healthy people. Tango dancing seems to be an appealing intervention regarding the body experience, however, publications are hardly available.

Question: Does tango dancing have an influence on the negative body experience of PD patients?

Method: Qualitative interviews (based on semi-structured guidelines) among 21 dance partners (with and without PD) who were participating Tango Argentine courses for persons with PD. The interview material was examined according to content analysis criteria (according to Mayring) and thematically coded (according to Flick).

Results: Five categories of body experiences were inductively generated during the analysis: 1. Body awareness (improved awareness of the body, i.e., movement and restricted movement), 2. Motor symptoms and movement (improved stability, walking safety, improvement in agility, gestures and facial expressions), 3. Body sensations (reduction of body fatigue, reduction of anxiety, shame and frustration, increase in joy, pride, curiosity and partner-related feelings), 4. General feelings (positive and negative feelings, ease, relaxation, increase of inner congruence), 5. Disease-related feelings and attitudes (experiencing normality and health, acceptance, development of self-confidence and self-confidence). Patients´ self-perception and external perception of their healthy dance partners were congruent.

Conclusion: Tango Argentine courses may have a positive influence on body experience by strengthening body awareness, body control and acceptance. During the dance, patients felt that their body was 'healthy' again and had pleasant body sensations that balanced their daily life's physical experience. In contrast to purely sporting interventions, Tango Argentine dance leads to a positive change on the level of subjective body experience, which has a therapeutic effect on the patients and their healthy partners.

In: Movement 2019: Brain, Body and Cognition
Editors: Joav Merrick and Gerry Leisman
ISBN: 978-1-53616-714-6

Chapter 116

TANGO ARGENTINE IN PATIENTS WITH PARKINSON'S DISEASE (PD). EFFECTS ON THE RELATIONSHIP EXPERIENCE FROM THE PERSPECTIVE OF THE HEALTHY DANCE PARTNER

Yvonne Beerenbrock[1], Stephanie Herrlich[2], Juliane Böhme[2], Lea Meyer[2], Sunnyi Mews[2], Bettina Berger[3], David Martin[3] and Arndt Büssing[1,*]

[1]Quality of Life, Spirituality and Coping, Institute for Integrative Medicine, Faculty of Health, Witten/Herdecke University, Herdecke, Germany
[2]Department of Psychology and Psychotherapy, Faculty of Health, Witten/Herdecke University, Herdecke, Germany
[3]Medical Theory, Integrative and Anthroposophic Medicine, Faculty of Health, Witten/Herdecke University, Herdecke, Germany

Keywords: Parkinson's disease (PD), Tango Argentine, relationship experience

Objective: Tango dancing not only improves the motor function of Parkinson's disease (PD) patients, but also promotes their quality of life and activity. The influence of tango on relationship experience has not yet been researched.

Question: Does the Tango Argentine have a perceptible effect on the relationship to the dance partner from the point of view of the (more objective) healthy dance partner?

Method: Qualitative interviews (based on semi-structured guidelines) of 10 healthy dance partners regarding partner perception. The text material was transcribed, evaluated according to content analysis criteria and grouped into categories.

Results: Eight main categories were derived: 1. partner perception (changed facial expressions, increase in physical stability), 2. self-esteem and self-efficacy expectation (increase in psycho-emotional stability of both dance partners), 3. relationship and leadership (changed leadership behaviour in dance as a resource in daily life), 4. experiencing physicality/intimacy (consciously experienced physical closeness/ appreciation), 5. relationship on communication level (promotion of non-verbal communication, communication of desires), 6. altered relationship level (rediscovery of healthy parts of partner and relationship), 7. feelings and sensations (positive/negative feelings) and 8. dealing with illness (new impulse from role distribution in dance for daily life).

Conclusion: Tango Argentine promotes inter- and intrapersonal aspects of dance partners with PD. From the perspective of the healthy partner, the quality of the relationship changes due to changed leadership roles, new forms of interaction, communication and common harmonious movement. Tango dancing as a therapy method touches all aspects of the disease-related relationship experience positively without overtaxing, but through pleasant and delightful harmonic partner movements with music.

In: Movement 2019: Brain, Body and Cognition
Editors: Joav Merrick and Gerry Leisman
ISBN: 978-1-53616-714-6

Chapter 117

THE EFFECTS OF TANGO DANCING ON THE MOTIVATION IN PATIENTS WITH PARKINSON'S DISEASE (PD)

Yvonne Beerenbrock[1], Juliane Böhme[2], Lea Meyer[2], Stephanie Herrlich[2], Sunnyi Mews[2], Bettina Berger[3], David Martin[3] and Arndt Büssing[1,*]

[1]Quality of Life, Spirituality and Coping, Institute for Integrative Medicine
Faculty of Health, Witten/Herdecke University, Herdecke, Germany
[2]Department of Psychology and Psychotherapy, Faculty of Health,
Witten/Herdecke University, Herdecke, Germany
[3]Medical Theory, Integrative and Anthroposophic Medicine, Faculty of Health,
Witten/Herdecke University, Herdecke, Germany

Keywords: Parkinson's disease (PD), Tango Argentine, motivation

Objective: Regular exercise plays a central role in delaying progressive mobility impairment in PD patients. Patients avoid regular movement for different reasons. For this reason, forms of movement that increase the motivation of those affected are important. The motivational dimension of tango dancing has not yet been adequately investigated.

Questions: How does tango dancing affect the motivational experience of PD patients?

Method: Qualitative interviews (based on semi-structured guidelines) of 11 patients and 10 unaffected dance partners who had taken part in Tango Argentine courses for patients. The text material was transcribed, evaluated according to content analysis criteria and combined into clusters.

Results: Four clusters can be generated from the interviews: A. motivational maintaining factors, B. routine transfer, C. demotivation and D. motivation before the course. The dance partners demonstrated a high degree of agreement in terms of content in the subcategories "shared joy" and "discovering new skills". Too high expectations of the intervention led to disappointment. For the PD patients, the discussion about the disease was of narrative relevance, while for the non-affected dance partners, their main motivation was for social contacts. Patients´ self-perception and external perception of their healthy dance partners were congruent.

Conclusion: Tango dancing motivates those affected by PD and their partners, in contrast to sporting interventions, by experiencing joint activities through the shared joy of joint creativity. The aspects of self-perception as well as the relational therapeutic dimension of tango dance are motivating for both partners. The central factor of the aesthetic experience through an experience of harmony has a therapeutic effect on the daily life of both partners and leads to a resource-oriented view of the illness as well as to social participation outside the course setting.

In: Movement 2019: Brain, Body and Cognition
Editors: Joav Merrick and Gerry Leisman
ISBN: 978-1-53616-714-6

Chapter 118

SUPPORTING EARLY INTERACTION BEHAVIORS THROUGH MOVEMENT PATTERN ANALYSIS IN CHILDREN WHO ARE CONGENITALLY DEAF-BLIND

Ivana Macokatic** and Ana Katušic
Day-care Centre for Rehabilitation Mali dom-Zagreb, Zagreb, Croatia and Croatian Institute for Brain Research, Centre of Excellence for Basic, Clinical and Translational Neuroscience, University of Zagreb, School of Medicine, Croatia

Keywords: early interaction, movement pattern analysis, congenitally deaf-blind children, communication

Objective: Early interaction behaviors with congenitally deaf-blind children are mostly supported by touch and proprioception. Both these qualities are essential elements of movements. In order to develop communication features such as turn taking, joint attention and dialogue it is important to identify movement patterns and qualities in children with deaf-blindness. The goal of this study is to investigate the development of early interaction behaviors in deaf-blind children in dance movement therapy setting.

Methods: Dance-movement therapy (DMT) sessions with 3 congenitally deaf-blind children up to 5 years of age were carried out once a week during 12 weeks. The Movement Pattern Analysis was applied to identify movement qualities in every child. The defined qualities were supported in DMT sessions during interactions in order to establish communication content. The video recordings of DMT sessions were performed initially, and at 6 and 12 weeks of intervention period. The frequency and duration of turn taking, joint attention and dialogue, were scored by two experienced therapists who were blinded to study questions.

Results: Results have shown significant prolongation of joint attention comparing to initial scores ($p < .04$). The frequency and duration of turn taking were also significantly increased after intervention period ($p < .02$). No differences were detected in dialogue behaviors.

Conclusion: Our study suggest that movement interpretation within dance-movement therapy intervention can enhance the development of early communication behavior in terms of turn taking and joint attention in congenitally deaf-blind children.

In: Movement 2019: Brain, Body and Cognition
Editors: Joav Merrick and Gerry Leisman
ISBN: 978-1-53616-714-6

Chapter 119

CORTICAL INTEGRATION OF CEREBELLAR SIGNALS DURING MOTOR LEARNING

Sharon Israely** and Yifat Prut
Department of Neurobiology, Edmond and Lily Safra Center for Brain Sciences,
The Hebrew University of Jerusalem, Jerusalem, Israel

Keywords: cortical integration, cerebellar signal, motor learning

Background: The cerebellum acts as a forward internal model that predicts the current sensory state, to allow calibration of motor output during movements. This mechanism is crucial to motor learning and adaptation. The cerebellum and the motor cortex interact via the cerebello-thalamo-cortical (CTC) pathway, which is assumed to be involved in relaying predictive signals from the cerebellum to the motor cortex. However, the nature and type of interactions between the cerebellum and motor cortex during error-based learning and the site where different types of adaptation are stored remain open questions.

Objectives: To study how cerebellar predictive signals are integrated by motor cortical neurons during motor adaptation, and how changes in performance during motor adaptation, when cerebellar output is interrupted, is correlated with cortical activity.

Methods: Two monkeys were trained to perform hand-reaching motor adaptation task to eight targets on the horizontal plane. A recording chamber was located above the motor cortex and a stimulating electrode was inserted into the Superior Cerebellar Peduncle (SCP). Motor performance during control trials and when the monkeys adapt to a novel force field were recorded. Low-frequency single-pulse stimulation, delivered through the SCP stimulating electrode, was used to identify motor cortical neurons that are part of the CTC system. A high-frequency stimulation (HFS) protocol was used to interrupt with the information flow through the CTC pathway. Acquired single cell activities were tested for significant responses to SCP stimulation. Changes in neural activity during adaptation trials were investigated to indicate for modulation in the motor command required for adaptation. Comparing these results to the changes occurring when adaptation is obtained in HFS trials highlighted the cerebellar contribution to these mechanisms.

Results: Response patterns to SCP stimulation were recorded from the primary motor (M1), premotor (PM) and somatosensory (SMC) cortices. High frequency SCP stimulation (HFS) efficiently blocked the flow of CTC information. This was shown by the complete loss of SCP-evoked response exhibited by cortical neurons. HFS impact on motor coordination, but did not prevent motor adaptation.

In: Movement 2019: Brain, Body and Cognition
Editors: Joav Merrick and Gerry Leisman
ISBN: 978-1-53616-714-6

Chapter 120

THE HEALING SPACE AND IMPLICIT LANGUAGE: THE NEUROLOGY OF THE SPACE "IN-BETWEEN" IN THE APPLICATION OF THE I-THOU RELATIONSHIP OF MARTIN BUBER

Vincent L Perri*
Private Practice Waterbury, Connecticut, United States of America

Keywords: space, in-between, neurology, Martin Buber

Objective: To explore and define the neurological underpinnings of Martin Buber's I-Thou relationship in the context of the practitioner-patient interaction, and the manner in which it facilitates inner healing in both physical and emotional disorders.

Methods: Numerous case studies of individual patients from the private practice scenario will be presented and evaluated in relation to the salience network and its role in the I-Thou relationship.

Results: Post therapeutic patient encounters will be reported and contrasted against their pre-therapeutic subjective complaints and evaluated in relation to the activation of specific neural pathways.

Conclusion: The I-Thou relationship of Martin Buber has deep and profound neurological underpinnings that can generate fundamental changes in both psychological and biological states. It can positively affect human behavior and has therapeutic effects in human healing.

In: Movement 2019: Brain, Body and Cognition
Editors: Joav Merrick and Gerry Leisman
ISBN: 978-1-53616-714-6

Chapter 121

COGNITIVE IMAGING OF MOVEMENT IN PERSISTENT VEGETATIVE STATES: BRAIN, BODY AND COGNITION

Calixto Machado* and Gerry Leisman
Institute for Neurology and Neurosurgery, Havana, Cuba
and University of Haifa, Haifa, Israel

Keywords: persistent vegetative state, movement, imagery, consciousness

Notwithstanding an astonishing assortment of scientific findings enhancing our understanding of the human body, we still know significantly little about the brain in general, and of consciousness in particular. Besides much debate over what constitutes a conscious mind, we are only now struggling with how our brains generate conscious awareness. The goal of this address is to attempt to identify a variety of neural signatures that can effectively indicate the presence of consciousness that could allow us to differentiate conscious and unconscious patients suffering from brain injuries in cases where individuals are unable to communicate. In the process of doing so, we have examined mental imagery of movement in patients in persistent vegetative and minimally conscious states.

While we do not know how it is that brain electrical activity is associated with consciousness, we are aware of the differences between conscious and unconscious brains. What we do know is that are consistent changes in the electroencephalographic output associated with various states of consciousness with attentive states having higher frequencies and lower amplitudes than progressively less conscious states. We also know that the brain-activity patterns in comatose and vegetative state individuals brains differ from those of the healthy individuals. Healthy, conscious brains demonstrate "rich and diversely connected networks," that are not found in the brain of comatose and vegetative state individuals.

The address will review what we know about the electrophysiology of consciousness and its disorders, the nature of inter-regional communication and the effects on the brain of the mental imagery of movement and its effect on both the motor system, cognition, and conscious awareness in individuals in persistent vegetative and minimally conscious states.

We will reference individuals who, despite being diagnosed as vegetative, with no capacity to perform voluntary physical movements in response to commands, when instructed to imagine performing a physical activity, such as playing tennis or kicking a football, the area of the brain responsible for controlling movement, became active. The vegetative patients demonstrating signs of hidden awareness have significantly well-preserved movement networks similar to those healthy adults". In addition, motor action words produce network motor system activations in language consistent for the individual as compared to arm, or face related words.

The new findings could help us identify "vegetative" patients who actually have some awareness, as well as improve their clinical assessment.

In: Movement 2019: Brain, Body and Cognition
Editors: Joav Merrick and Gerry Leisman
ISBN: 978-1-53616-714-6

Chapter 122

BIOMECHANICAL BASIS OF TAI CHI MOVEMENTS FOR INTERVENTION OF LOW BACK PAIN PATIENTS ASSOCIATED WITH MUSCLE DISORDERS: A PILOT STUDY

***Jing Xian Li*[1,*], *Jian Kong*[2], *Xin Zhou*[3], *Qingguang Zhu*[3] *and Ming Fang*[3]**

[1]School of Human Kinetics, University of Ottawa, Ottawa, Canada,
[2]Department of Psychiatry, Massachusetts General Hospital, Harvard Medical School, Charlestown, Massachusetts, United States of America
[3]Tui-na Research Department, Shanghai University of Traditional Chinese Medicine, Shanghai, China

Keywords: Tai Chi, low back pain, gluteus medium, electromyography

Objective: This study examined the muscle activity of the trunk and lower limb in order to provide scientific basis for developing a biomechanical based Tai Chi intervention program for management of the low back pain (LBP) related to muscle disorders in the patients.

Method: One experienced Tai Chi master with more than ten years Tai Chi experiences participated in the study. Four Tai Chi movements are selected for analysis, including 1) wave hand in cloud, 2) brush knee twist step, 3) repulse monkey, and 4) leaning fly side. A motion analysis system (Vicon, Oxford, UK) with sampling rate of 200 Hz and three force plates with sample rate 1000Hz were used to record the master's Tai Chi movements and ground reaction forces (GRF). EMG signals from right rectus femoris, adductor longus, tibialis anterior, biceps femoris, semitendinosus, erector spinae, gluteus medius, tensor fasciae latae, medial and lateral ganstrocnemius muscle were collected during performing Tai Chi using surface Wireless EMG System (SP-W01D, Delsys Inc., Boston, MA). Tai Chi gait cycle was defined from foot strike to foot strike for the same foot. Heel strike and toe off were defined by GRF and by visual inspection of the position of the virtual heel and toe markers of motion pictures. Raw EMG signals were processed using custom MatLab program. The processed EMG data were normalized to 100% of the Tai Chi gait cycle and reported as maximal iEMG and EMG duration in a gait cycle.

Results: All lower limb muscles, particularly gluteus medium (GM) and tensor fasciae latae (TFL) that have been linked to LBP, showed significant muscle activation patterns during Tai Chi movements. The EMG duration of the GM and TFL in one gait cycle accounts for 52% and 69% of a gait cycle time respectively.

Conclusion: EMG activity patters of the lower limb and trunk muscle, particularly GM and TFL, during performing selected Tai Chi movements suggest that the Tai Chi movements would be suitable for training the muscle function for the patients with LBP associated with muscle disorders of hip and trunk.

In: Movement 2019: Brain, Body and Cognition
Editors: Joav Merrick and Gerry Leisman ISBN: 978-1-53616-714-6

Chapter 123

EAST AND WEST: TOWARDS AN INTEGRATIVE GRAND UNIFIED THERAPY (GUT)

David Barend Labuschagne*
ENE (Energijska Znanost in Umetnost), Izola, Slovenia

Keywords: mind-body therapies, chronic disease, aging, Taiji-Qigong, falls prevention

Objective: Mind-body therapies such as Qigong, Tai Chi and Yoga are gaining acceptance in the fields of rehabilitation and health conditioning. The author has developed an innovative modality, Kinergy Movement Therapy (KMT), integrating Western exercise, human-movement and physical therapy sciences with Eastern mind-body therapies and martial arts training. Long-term trials have been conducted to assess KMT's potential for chronic disease management, falls prevention, and the amelioration of aging-related health issues.

Methods: KMT programs consist of weekly 90-minute group sessions staged over 13 weeks. Each session comprises a 60-minute workout supplemented with a 30-minute lecture on health topics. KMT trials were staged over a 12-year period with a cohort of 470 participants presenting with a broad range of chronic and aging-related pathologies. Cohort age range was 18-86 years, median age 52 years. Program participation was prefaced by, and concluded with interviews; assessment of fitness and physical functionality; medical history review; and pathology reports. These factors along with client self-reports were used to evaluate outcomes. Long-term outcomes were evaluated in follow-up programs, private sessions and interviews, and extended over a period of 5 years from participants' enrolment dates.

Results: KMT was productive of significant and enduring outcomes in the management of a broad range of chronic, complex and aging-related pathologies. KMT was also effective in managing psychological issues such as anxiety, depression and self-efficacy. Compliance rate exceeded 90%, and virtually all participants reported and displayed significant improvements in mobility, vitality and functional performance. No adverse events were reported.

Conclusion: KMT has demonstrated substantial potential for delivering significant and enduring health benefits to the broad demographic at risk of, or afflicted with chronic and aging-related health issues. In addition to symptom relief, the benefits include increased functional capacity; a healthier, more independent lifestyle; and extension of productive lifespan. KMT programs and program-leader training courses are scalable and suitable for delivery by a range of healthcare providers as well as transmission via digital media formats. Further considerations of safety of implementation, cost-effectiveness of group sessions, minimal requirement for specialised facilities or ancillary aids, and outstanding participant compliance signifies the importance of wider implementation and dissemination.

In: Movement 2019: Brain, Body and Cognition
Editors: Joav Merrick and Gerry Leisman

ISBN: 978-1-53616-714-6

Chapter 124

DANCE PARTICIPATION STABILIZES MOTOR AND COGNITIVE SYMPTOMS OF PARKINSON'S DISEASE AND AGE-RELATED DISEASES

***Rebecca Barnstaple*[1-3,*], *Karolina Bearss*[1] and *Joseph FX DeSouza*[1,3-5,†]**

[1]Centre for Vision Research, York University, Toronto, Canada,
[2]Department of Dance,
[3]Neuroscience Graduate Diploma Program,
[4]Canadian Action and Perception Network (CAPnet)
[5]Departments of Psychology and Biology, York University, Toronto, Canada

Keywords: dance, Parkinson's disease, UPDRS, QoL, life space

Objective: Parkinson's disease (PD) and other age-related diseases are characterized by debilitating and progressive symptoms, costing the health care system billions of dollars annually. Our research program indicates that a cost-effective weekly dance program is highly effective in reducing symptoms associated with PD and other age-related diseases. Our research aims to demonstrate how dance participation can support healthy aging through 1) increased quality of life, 2) slowed disease progression 3) reduced mental health comorbidities (depression/anxiety), 4) reduced falls, 5) reduced reliance on medication and 6) expansion of the "life space" and success in activities of

[*] Corresponding Author's Email: rebecca.barnstaple@gmail.com
[†] Corresponding Author's Email: desouza@yorku.ca.

daily living (ADL). In this abstract we present results from items 1-3 and discuss the rationale for future/continued research in 4-6.

Method: Subjects (N = 102) composed of people with PD (N = 50) and healthy controls (N = 42) participate in a 1.25-hour dance program 1X/week. Measures include the United Parkinson's Disease Rating Scale (UPDRS), Mini-Mental State Exam (MMSE), Geriatric Depression Scale (GDS), Quality of Life (QoL), Berg Balance Scale (BBS), Timed up and go (TUG), and rsEEG.

Result: In relation to the above aims, thus far our research program has shown:

- QoL improvements
- Longitudinal changes in people with PD where disease progression is halted in a small cohort
- Improvements in mental health examined in GDS as a function of attending regular dance class and across time.

Items 4-6 outline the direction of current and future research.

Conclusion: Dance, a highly cost-effective intervention, can improve function and quality of life for people with progressive and pervasive disorders, without negative side-effects. Dance programs reduce social isolation for participants, moving them out of their habitual housing to a larger space, pushing their cognitive maps and increasing Life Space. This in turn supports better/continued performance in ADL, directly linked to independent living. Continued research can further demonstrate the benefits and effectiveness of dance-based programs, which have been undervalued in terms of their potential to contribute to healthy aging and a reduction in the global burden of disease.

In: Movement 2019: Brain, Body and Cognition
Editors: Joav Merrick and Gerry Leisman

ISBN: 978-1-53616-714-6

Chapter 125

LOCOMOTION AND MOVEMENT COMPLEXITY

David Gallina*
David Gallina Movement Education, LLC and Bastyr University in Seattle, Washington, United States of America

Keywords: functional neurology, movement, development, therapy

This movement-based workshop was presented at the International Association of Functional Neurology and Rehabilitation's 2018 annual conference. We will explore various patterns of locomotion - from bipedal to quadrupedal movements as well as options for communicating with the floor. This material can be scaled from neurological rehab settings to optimal athletic performance and is aimed to benefit both you and your patients. We can adjust variables of this material to develop aspects of motor planning and control, coordination, balance, rhythm, proprioception, strength, midline stability, joint health/mobility, metabolic capacity, and minimizing fall risk. Please wear clothes you can move in.

In: Movement 2019: Brain, Body and Cognition
Editors: Joav Merrick and Gerry Leisman
ISBN: 978-1-53616-714-6

Chapter 126

MONOCULAR AND BINOCULAR OBSERVATIONS AND DOMINANT AND NON-DOMINANT EYE OBSERVATIONS CAN MODULATE THE PERCEIVED STRENGTHS OF VISUALLY INDUCED SELF-MOTION (VECTION)

Takeharu Seno * ***and Takaya Shimizu***
Faculty of Design, Kyushu University and Kyushu Economic Research Center, Kyushu, Japan

Keywords: vection, monocular, binocular, dominant eye, eyepatch

Objective: When a large visual filed is occupied by a coherent motion, the observers would perceive their own body motion illusorily. This visually induced self-motion perception is named vection. We examined whether vection strength could be modulated by the difference in the conditions of observing eyes, i.e., monocular or binocular observations and also, dominant eye or non-dominant eye observations.

Method: The eyepatch was used for making the monocular observation. To check which eye was the dominant eye was conducted by Miles method. Eleven naïve volunteers and the second author participated in this experiment. The average age was 23.9 years old (SD = 4.7). Three females and nine males participated. All of them had normal vision. Three vection measures (latency, duration and magnitude) were obtained. These three indices have been obtained in a lot of previous vection studies. As a visual stimulus, we presented the participants an optical flow stimulus, i.e., expanding dots' motion. The duration of the stimulus presentation was 40 seconds. There were three conditions, binocular, left eye and right eye observation conditions.

Results: Obtained vection latency was significantly longer in the left eye observing condition, although there were no significant differences in vection duration and magnitude between the three conditions. Shorter latency suggests that perceived vection strength was weaker when the stimulus was observed by the left eye. The left eye was the non-dominant eye in nine out of twelve participants. Thus, we might be able to say that the non-dominant eye observation can weaken the perceived vection.

Conclusion: We can conclude that vection can be affected by the conditions of observing eyes. The non-dominant eye observation might weaken the perceived vection. This knowledge should be used in some future VR environments.

In: Movement 2019: Brain, Body and Cognition
Editors: Joav Merrick and Gerry Leisman
ISBN: 978-1-53616-714-6

Chapter 127

A CHALLENGE TO MAKE MOVIES "WHAT IS VECTION!?" AND THE EVALUATIONS OF THEM BY SEVEN PROFESSIONAL CREATORS OF VISUAL IMAGES

Tomohiko Akagi*, Richika Urasoko, Tokushu Inamura and Takeharu Seno†
Faculty of Design, Kyushu University, Kyushu, Japan

Keywords: vection, movies, professional creators, visual images

Objective: When a large visual filed is occupied by a coherent visual motion, the observer would perceive his/her own body motion in the opposite direction to the visual motion. This visually induced illusory self-motion perception is named "Vection". Vection has been studied for about 40 years in behavioral and cognitive psychology. Setoguchi and Seno made various types of vection movies that were based on over 30 psychological published articles. Also, we checked many Japanese animation films and found 150 vection scenes. We uploaded the information of them on our web site (senotake.jp). Our challenge is to make much more people know "vection". In this present study, we newly made some movies in which lectures of vection were conducted and then we made seven professional creators of visual images evaluate these movies and our approaches.

* Corresponding Author's Email: akgtmhk0317@gmail.com.
† Corresponding Author's Email: seno@design.kyushu-u.ac.jp.

Method: In the movies, what vection is, was fully explained scientifically. The history of vection and "Vection and virtual reality" were interestingly introduced in the movies. The languages used in these movies were Japanese and English. Secondary, we interviewed seven professional creators of visual images, on these vection movies. Seven professional creators evaluated our vection movies. In the seven, three contemporary artists, one movie director, two designers and one photographer were included.

Results: We got a lot of hints for the further developments of our approaches from the results of these interviews. All of the seven creators recognized the value of vection as a visual effect in various possible contents. They suggested the future tasks for our approaches to bridge the two different business fields. They said to us that the efforts of experimental psychologists are not enough so far. They asked us to make much more efforts to make vection much more popular.

Conclusion: We sincerely hope that this study can be the bridge between experimental psychologists and contents creators and developers. Finally, we want to make vection itself as ART!

In: Movement 2019: Brain, Body and Cognition
Editors: Joav Merrick and Gerry Leisman
ISBN: 978-1-53616-714-6

Chapter 128

KINESIO TAPING CAN DO HARM IN PATIENTS WITH TYPE 1 COMPLEX REGIONAL PAIN SYNDROME (CRPS-1) AFTER STROKE

Xian Li**, *Zhi Zou, Tao Zhou, Xin Wang and Yuling Wang
The Sixth Affiliated Hospital of Sun Yat-sen University, Guang Zhou, China

Keywords: CRPS-1, kinesio taping, LRP, aging

Objective: This study aims to investigate the effect of Kinesio taping in patients with type 1 complex regional pain syndrome (CRPS-1) after stroke.

Methods: Seventeen patients with CRPS-1 after stroke were randomized to the taping group and control group. The control group received standard treatment, including the active and passive range of motion, spasticity management, motor relearning, weight bearing exercise, etc., while the taping group received standard treatment plus Kinesio Taping. However, modality that aim to reduce edema and pain were excluded for both groups. Treatments were delivered for two weeks. The edema of the affected hand was evaluated as the primary outcome and the pain level and the physical function were measured as the secondary outcome. The pain was evaluated by Visual Analogue Scale and the physical function was measured by Upper Limb portion of the Fugl-Meyer Assessment and pain-free lateral rotation range of shoulder. Two-way repeated ANOVA was used in the statistical analysis. We evaluate the participants at baseline and two week post intervention.

Results: The edema of the affected hand evaluated by Hand Volumeter showed no interaction effect between two groups ($P = 0.297$). However, the control group showed significant decrease of hand edema ($P = 0.003$) while the volume of the affected hand increased 11ml in the tapping group (P = 0.389). Similarly, no significant interaction effect was observed in the pain assessment (P = 0.163) as well as the physical function measurement, including the Fugl-Meyer Assessment (P = 0.163) and pain-free lateral rotation range of shoulder (P = 0.640).

Conclusion: The taping for post-stroke patients with CRPS-1 could not decrease the pain level or increase the function level of the stroke patients with CRPS-1. Moreover, the tapping could increase the hand edema level. These results indicated that although the patients with musculoskeletal disorder could benefit from taping, this treatment could do harm to the patients with CRPS-1 after stroke. This kind of harm could be related to the different mechanism of edema. Thus, for the edema caused by neurological disorder may not be suitable for tapping.

In: Movement 2019: Brain, Body and Cognition
Editors: Joav Merrick and Gerry Leisman
ISBN: 978-1-53616-714-6

Chapter 129

THE PERCEPTION OF OUR MOVEMENTS DEPENDS ON OUR EMOTIONAL STATE

Edith Ribot-Ciscar**, *Léonard Samain-Aupic, Jean-Marc Aimonetti and Rochelle Ackerley

Sensory and Cognitive Neuroscience Laboratory, Aix Marseille University, Marseille, France

Keywords: movement sense, proprioception, muscle afferents, kinesthesia, emotion

Objective: The perception of our movements relies on the integration of proprioceptive information arising from sensory receptors in our muscles (the muscle afferents). The aim of the present study was to explore whether our emotional state alters the muscle sensory messages and affects the perception of limb movements.

Methods: Neurophysiologic approach: The unitary activity of muscle afferents to movements was recorded at the level of the common peroneal nerve by microneurography. Their response to movements of the ankle was compared while the participant listened to evocative classical music to induce sad, neutral or happy feelings. Psychophysical approach: Kinesthetic acuity was tested by imposing very slow movements that consisted of either plantar flexion or dorsiflexion of the ankle in the same emotional contexts.

Results: Emotion can alter the proprioceptive messages from muscle afferents, making them more sensitive to movements when participants were listening sad music. Kinesthetic acuity was increased during the sad condition, as compared to the no music or neutral conditions.

Conclusion: We show that the manner in which our muscles inform our brain about movements changes depending on our emotional state. This influence of cognitive mechanisms on muscle proprioceptive information has direct consequences on our perception of body movements.

In: Movement 2019: Brain, Body and Cognition
Editors: Joav Merrick and Gerry Leisman
ISBN: 978-1-53616-714-6

Chapter 130

THE NEUROMODULATIVE EFFECTS OF TIREDNESS AND MENTAL FATIGUE ON COGNITION AND THE USE OF MEDICATION

José León-Carrión[1,*], Umberto León-Dominguez[2] and Maria del Rosario Dominguez-Morales[3]
[1]Department of Experimental Psychology, University of Seville, Seville, Spain
[2]Department of Health Services, School of Psychology, University of Monterrey, Monterrey, Mexico
[3]Center for Brain Injury Rehabilitation (CRECER), Seville, Spain

Keywords: mental fatigue, tiredness, neuropsychology, dopamine, hypnotic drugs

Objective: Fatigue is defined by means of subjective and neurophysiological deficits that vary as a function of time and workload. Cognitive load and task difficulty mediate fatigue. Performance deficits normally demonstrate evidence of fatigue and awareness of mental effort, although overt deficits may not be a consequence of fatigue, as increased efforts normally compensate the decline of mental resources. Subjective mental fatigue leads to increased mental effort, impaired learning and stress. Homeostatic and circadian components of fatigue are different neurophysiological mechanisms affecting mental fatigue. One of them is the dopamine system.

Methods: Coping with the effects of fatigue often require pharmaceutical measures to maintain, or even restore, operative cognitive functions. The more relevant aspects of this presentation are related to the following points: homeostatic and circadian components of fatigue, fatigue and tiredness management strategies, the use of hypnotic and other drugs, short and long-term efficacy of drugs and recommendations. One of them is the dopamine system. Mental fatigue provokes significant changes in the brain, altering the glutamate network, particularly in challenging situations, including chronic pain and neurodegenerative disease. Mental work is compromised by mental fatigue Legal and illegal drugs also may improve cognitive task performance over an extended period. Mental fatigue is reversed simply by resting or taking time away from the task. Sleep is normally restorative. A short nap after hard work reduces the effects of fatigue.

Results: People coping with the effects of fatigue must rely on pharmaceutical measures if they need to maintain, or even restore, operative cognitive functions. The most recommended pharmaceutical measures to combat fatigue are: Provigil, hypnotics, tryptophan, zinc, magnesium, vitamin B and mutli-vitamin complex.

Conclusion: Recovery from fatigue is linked to the control of perseverance, consistency and efficacy during a specific period. Fatigue or tiredness is the logical consequence of carrying out a mental and/or physically task requiring perseverance, consistency and efficacy during a specific period. Recovery from fatigue is linked to the control of these three factors.

In: Movement 2019: Brain, Body and Cognition
Editors: Joav Merrick and Gerry Leisman
ISBN: 978-1-53616-714-6

Chapter 131

BODY PERCEPTION IN STROKE SURVIVORS: A PILOT STUDY ON THE EFFECT OF EXPRESSIVE ARTS-BASED INTERVENTION

Temmy Lee Ting Lo[1,*] and Rainbow Tin Hung Ho[1,2]
[1]Department of Social Work and Social Administration,
The University of Hong Kong, Hong Kong
[2]Centre on Behavioral Health, The University of Hong Kong, Hong Kong

Keywords: stroke, expressive arts, somatics, body perception, rehabilitation

Introduction: The traditional rehabilitation program mostly focuses on stroke survivor's physiological or biological dysfunctions and symptoms. The subjective perception towards one's body is rarely discussed, yet this may be vital in the post-stroke rehabilitation trajectory and psychological conditions. Accordingly, this pilot study applied the "inside-out", subjective view of a body, which advocates in the somatic field. By adopting Expressive Arts-based Intervention, this study aims at exploring how this intervention changes stroke survivors' first-person body perception and psychological status.

Methods: Participants were recruited in a local rehabilitation center. Eligible participants, with age above 18 years old and with a major stroke event within five years, were referred by the center's staff. Participants joined the eight-weeks Expressive Arts-based Intervention group which applies multiple art modalities for creating multiple sensory stimuli and facilitates participants to process and express the complex feelings towards their bodies. In-depth interviews supplemented with drawing assessment were also conducted to gain more understanding of the differences in their subjective body perception. The quantitative assessments were also included to assess the level of hope, anxiety, depression, and perceived stress.

Results: Six participants, aged 44 to 70, participated in this study. When comparing the sketches they drew before and after the intervention, the use of colours was more harmonious. Moreover, the post-intervention sketches tend to have more details. This may present that participants gained more body awareness and experienced improvements in body coordination. During the interviews, participants indicated that the Expressive Arts-based Intervention enhanced their mood and relieved their anxiety. There were also physical improvements through routine rehabilitation sessions. Those changes maybe related to the differences in the sketches. For the quantitative measurements, there was also an increased trend of hope level and decreased trend of anxiety, depression, and perceived stress level.

Conclusion: This pilot study provided initial findings that Expressive Arts-based Intervention may change stroke survivor's first-body perception, induce hope, and lower the level of depression, anxiety, and perceived stress. While during the interviews, some participants mentioned the mood enhancement may be related to the changes in first-person body perception, more evidence is needed to confirm this relationship.

In: Movement 2019: Brain, Body and Cognition
Editors: Joav Merrick and Gerry Leisman
ISBN: 978-1-53616-714-6

Chapter 132

Multidisciplinary intensive outpatient rehabilitation program for patients with advanced Parkinson's disease: Feasibility and primary proof of concept

***Tanya Gurevich*[1,2], *Ytzhak Green*[3], *Noit Inbar*[1], *Geyl Tahel*[3], *Noa Cohen*[3,*], *Omer Kadurri*[3], *Liat Shtraifler*[3], *Inbal Badichi*[3], *Neta Rips Casseri*[3], *Ariella Hilel*[1], *Norit Gheriani*[1], *Adi Ezra*[1], *Stav Naor*[1], *Yami Shapiro*[1], *Nir Giladi*[1,2] *and Yael Manor Yael*[1,4]**
[1]Movement Disorders Unit, Neurological Institute, Tel-Aviv Medical Center, Tel Aviv
[2]Sackler School of Medicine, Sagol School of Neuroscience, Tel-Aviv University, Tel-Aviv
[3]Ezra Lemarpe Rehabilitation Center, Bnei-Brak
[4]Ono Academic College, Kiryat Ono, Israel

Keywords: advanced Parkinson disease, rehabilitation program

Background: Intensive, multi-dimensional rehabilitation programs for patients with Parkinson's disease (PD) have been shown to be effective. However, most programs are based on in-patient service which is extremely expensive and can be provided for a relatively short period of time.

Objective: To describe as a proof of concept our experience with a multidisciplinary, intensive out-patient rehabilitation program (MIOR) for advanced PD patients.

Method: The MIOR is an 8 week (3 days a week, 5 hours per day), free of charge program, which takes places in a single community-based rehabilitation center. A wide range of multidisciplinary treatments are offered, including physical occupational and speech therapy and consultation from a social worker and rehabilitation neuropsychologist who meet all patients and spouses. The healthcare team provides individualized treatment, alongside treatment for the entire group or sub-groups, according to gender, disease severity or specific disabilities. Additional physical and social activities include hydrotherapy, boxing, dancing, singing and Bridge (card game), as well as relaxation sessions.

Results: Since June 2016, three groups (20 participants in each group) have attended the intense rehabilitative program: 53 participants with advanced PD (H&Y = 3) completed the full program (120 hours); 7 patients were unable to complete the program due to health-related issues. Over all, patients and spouses have expressed great satisfaction from the program and most participants reported significant improvements in their daily function which could be attributed to the program. Objective measures will be presented.

Conclusion: Based on 6 months of experience with 3 groups, we can conclude that MIOR is an achievable health-promoting multidisciplinary rehabilitative model for people with PD. Most patients have shown motivation, active participation and adherence throughout the program. Therefore, it could be feasible and significant for patients with PD, especially if covered by health insurances. Long-term outcomes should be reported periodically using objective measures, in order to persuade the health authorities to cover such an expensive yet beneficial program.

In: Movement 2019: Brain, Body and Cognition
Editors: Joav Merrick and Gerry Leisman
ISBN: 978-1-53616-714-6

Chapter 133

Multidisciplinary group program integrating voice and dance movement therapy for Parkinson's disease patients: Preliminary experience

Roni Peled[1], Dina Shpunt[1], Yael Manor[2], Marina Brozgol[3,4], Adi Ezra[1], Jeff Hausdorff[3,4] and Tanya Gurevich[1,*]
[1]Movement Disorders Unit, Department of Neurology,
Tel Aviv Sourasky Medical Center, Tel Aviv, Israel
[2]Ono Academic College, Faculty of Health Professions,
Communication Sciences and Disorders Department, Kiryat Ono
[3]Sackler School of Medicine and Sagol School of Neuroscience
Tel Aviv University, Tel Aviv, Israel
[4]Center for the Study of Movement, Cognition, and Mobility, Department of Neurology,
Tel Aviv Sourasky Medical Center, Tel Aviv, Israel

Keywords: Parkinson's disease, voice, motion, quality of life

Background: Physical exercise and speech therapy are effective treatments for Parkinson's disease (PD). The "Voice in Motion" group activity is led by movement and speech therapists. It encourages patients to perform speech drills based on LSVT-LOUD and spontaneous speech and dance movement therapy (DMT), while at the same time emphasizing posture, balance and lively interactions within the group.

Objective: To examine the effects of "Voice in Motion" group therapy on movement, speech and QOL of PD patients.

Method: The parameters of volume of vowel prolongation of/aaa/in a habitual (VVPh) and loud (VVPl) voice in spontaneous speech (VVSS), gait speed and stride length and responding to questionnaires voice handicapped index (VHI), Visual Analog Scale (VAS), gait parameters and Quality of Life (PDQ-39) were compared before and after 27 weekly sessions in this pilot retrospective report.

Results: Six patients (two males), mean age 64.5 ± 7.8, Hoen and Yahr 2.5 ± 0.6; disease duration 3.7 ± 2.7 and Montreal Cognitive Assessment (MoCA) 26.8 ± 1.7 participated in 27 weekly sessions. A positive trend towards improvement was noted in all measurements: VVPH, VVPl and VVSS (T1 68.2 ± 10.5dB, T2 80.8 ± 4.9dB, P = 0.068; T1 89.5 ± 5.4dB, T2 96 ± 6.6dB, P = 0.279; T1 60.3 ± 4.8dB, T2 65.2 ± 5.4dB, P = 0.131, respectfully), and gait speed (T1 0.16 ± 1.01 m/sec, T2 1.06 ± 0.3 m/sec; P = 0.73). The VHI, emotional category (T1 7 ± 8.3, T2 4.2 ± .2, P = 0.197) PDQ-39, stigma parameter (T1 3.2 ± 2.7, T2 2.7 ± 2.3, P = 1.000) Patients reported: "The group gave me self-confidence to discuss the disease with family and friends. No falls and adverse events were reported.

Conclusion: Therapy that combines speech and movement is apparently safe and feasible. Initial results suggest that integrating voice and dance movement therapy that focuses on expanding the use of bodily expression and social communication can maintain and improve speech, movements and communication abilities and well-being of PD patients.`

In: Movement 2019: Brain, Body and Cognition
Editors: Joav Merrick and Gerry Leisman
ISBN: 978-1-53616-714-6

Chapter 134

A MoCo-INTERVENTION REDUCES THE RELIANCE ON CORTICAL RESOURCES DURING PHYSICAL TASKS OF DAILY LIVING

Dennis Hamacher*[1,*]*, Kim-Charline Broscheid*[1]*,
***Fabian Herold*[2] *and Lutz Schega*[1]**
[1]Department III, Sport Science
[2]German Centre for Neurodegenerative Diseases, Otto von Guericke University Magdeburg, Magdeburg, Germany

Keywords: fNIRS, gait, automaticity

Objective: Walking safely requires the activation of specific cortical structures, particularly the dorsolateral prefrontal cortex (DLPFC) and the supplementary motor cortex (SMA) (Hamacher et al., 2015). Furthermore, while emerging evidence suggests that the amount of cortical control needed for executing walking tasks is moderated by individual motor-cognitive abilities, there is little evidence available how different types of physical interventions influence cortical control during walking. Hence, the purpose of the study is to compare the effects of two different physical intervention strategies on the activation of in DLPFC and SMA while walking.

Methods: 25 older subjects were randomly assigned to either a health-related exercise group (Control: age = 69.9 ± 4.3 years) or a multi-modal motor-cognitive intervention group (MoCo: age = 70.2 ± 5.8 years) exercising for ten weeks. In a pre-test and post-test, we investigated the cortical activation (objectified via oxygenated [oxyHb] and deoxygenated hemoglobin [deoxyHb]) during overground dual-task walking by using a portable functional near-infrared spectroscopy system. Data were analyzed using two-way repeated measures ANOVAs (time by group).

Results: We found main time effects from the pre to post test in the right DLPFC (oxyHb: $p = .017$; deoxyHb: $p = .004$) and the right SMA (oxyHb: $p = .009$; deoxyHb: $p = .020$). Also, an interaction effect was found regarding deoxyHb of the right SMA ($p = 0.033$) in favor of the MoCo group.

Conclusion: Our data indicates that the multi-modal intervention strategy MoCo combining motor and cognitive efforts reduces the reliance on cortical resources during dual-task walking. This, in turn, may lower fall risk in challenging physical activities of daily living with divided attention. Thus, engaging in a multi-modal motor-cognitive training might be an efficient way to prevent future falls.

In: Movement 2019: Brain, Body and Cognition
Editors: Joav Merrick and Gerry Leisman
ISBN: 978-1-53616-714-6

Chapter 135

FUNCTIONAL NEUROLOGY IN A NUTSHELL: A MODEL FOR SCIENTIFIC UNDERSTANDING OF NERVOUS SYSTEM FUNCTION AND CLINICAL APPLICATION

Robert Melillo*

Department of Physiotherapy, Faculty of Health Sciences, University of Haifa, Haifa, Israel and The National Institute for Brain and Rehabilitation Sciences, Gilbert, Arizona, United States of America

Keywords: functional neurology, brain networks, neuroplasticity

Functional Neurology is both a clinical term and a professional discipline. It has been coined as a clinical term in the 1970's and has been used in recent years in the world of medical neurology to address interventions aimed at increasing overall function, such as in Parkinson's and other neurodegenerative disorders. As a discipline, it began as a specialty of Chiropractic, but has since be adopted by other health care professions. It is based on the use of receptor-based applications to change brain function based on a fundamental understanding of how neuroplasticity may be affected by external means.

Research has demonstrated much about the nature of the development and change in network functions of the brain, we have found that every function has a "lead" region, such as the dorsolateral prefrontal cortex in executive function. This allows a practitioner to examine the nervous system and understand the adequacy of function of brain regions associated with an individual's symptoms. Based on an understanding of laboratory-based understanding of how neuroplasticity may be affected, changing the sensory input to a certain region, or inducing motor/cognitive functions associated with that region, can shunt blood flow to that region, and induce cellular and intercellular changes based on immediate and long-term responses that will result in a change in its function. This, it is hypothesized; will as a result, change functional output, which we can now effectively measure. This approach requires the practitioner to be familiar with anything that may affect brain function including nutrition, electrical and electronic modalities, pharmacology, exercise, manual therapies and more. This versatility is most effective when employed as part of a multidisciplinary approach, and can provide possible solutions for many conditions once though incurable, decreasing and even reversing symptoms, increasing speed of rehabilitation, improving performance etc. The integrative nature of Functional Neurology as an intervention strategy rests on the nervous systems integrative capacity most exemplified by the relation between movement and cognition.

In: Movement 2019: Brain, Body and Cognition
Editors: Joav Merrick and Gerry Leisman
ISBN: 978-1-53616-714-6

Chapter 136

THE USE OF UNILATERAL SENSORY/MOTOR STIMULATION TO INHIBIT PRIMITIVE REFLEXES AND PROMOTE HEMISPHERIC INTEGRATION IN CHILDREN, ADOLESCENTS AND ADULTS WITH NEUROBEHAVIORAL MENTAL HEALTH DISORDERS

Robert Melillo*
University of Haifa, Haifa, Israel and The National Institute for Brain and Rehabilitation Sciences, Gilbert, Arizona, United States of America

Keywords: functional neurology, rehabilitation, child development, primitive reflexes, hemisphere-specific training, immune dysregulation

Developmental functional neurology, is based on the principle that what happens developmentally in the first six years of life may have an impact on the long-term development and function of the brain and subsequent behavior. Recent statistics show that 20% or 1 in 5 children and a similar number of adults will be diagnosed with a severe mental disorder in the USA. We propose that children with ADHD, Autism Spectrum Disorder, or related diagnosis as adults manifest these same or similar issues. It is well documented that Schizophrenia may not manifest until early adulthood, but that it starts in the womb and may have some early clues that affect motor developmental milestones within the first few years of life. Since most treatment modalities do not address the root causes of these issues, do not have long-term benefit and that individuals do not generally "outgrow" these disorders, is reasonable to assume that if a child is diagnosed with a mental health issue that issue will remain with them for life. Therefore, if we can identify and treat the root causes of these behavioral dysfunction at an early enough point in an individual's development, can we then not only eliminate or reduce the childhood disorder but also prevent its adult manifestations. This aspect of the keynote symposium on Functional Neurology proposes that examining the retention of primitive reflexes in children, adolescents and adults is an effective means of identifying developmental neurological dysfunction and one objective marker for evaluating improvement of any treatment or intervention. Children within the first year of life should be able to inhibit and integrate primitive reflexes and release postural reflexes allowing normal motor and cognitive development. If these motor primitives are not inhibited or effectively bilaterally integrated, then it is more likely to find an associated asymmetry of tone, in turn associated with altered motor development and developmental milestone delays. This asymmetry of tone and movement is further associated with developmental asynchrony between developing networks between the brainstem and the neocortex. The associated thalamocortical dysrhythmia is ultimately associated with a functional disconnection between the cerebral hemispheres. We hypothesize that this is the root cause of many neurobehavioral disorders and mental health issues in children and adults. This also can lead to immune and autonomic dysregulation, which can have a significant impact on an individual's general health status. We have found that one of the most effective ways of inhibiting primitive reflexes is through the use unilateral sensory stimulation directed toward the underdeveloped hemisphere. Light, sound, tactile (TENS), smell, vestibular, optokinetic and motor activity can be used in various combinations to reduce or completely eliminate primitive reflexes at any age and reduce cognitive and physical symptoms as well.

In: Movement 2019: Brain, Body and Cognition
Editors: Joav Merrick and Gerry Leisman
ISBN: 978-1-53616-714-6

Chapter 137

THE IMPACT OF INTENSIVE MULTIMODAL FUNCTIONAL NEUROLOGICAL APPLICATIONS ON COGNITIVE OUTCOMES IN A POPULATION WITH POST-CONCUSSION SYNDROME AND ASSOCIATED LEARNING AND BEHAVIORAL DISORDERS

Michael S Trayford*
APEX Brain Centers, Asheville, North Carolina, United States of America

Keywords: functional neurology, TBI, concussion, ADHD, addiction

Objective: This presentation seeks to demonstrate a viable link between improvements in motor timing, vestibular integration, oculomotor efficiency, motor skills, core/postural stability, and brainwave regulation; and the positive impact of such on specific cognitive domains in a small population of individuals with post-concussion syndrome.

Methods: A sample of ten individuals was chosen from our database with concussion symptoms lasting more than six weeks, or post-concussion syndrome (PCS) as the primary diagnosis; and attention deficit-hyperactivity disorder (ADHD), obsessive-compulsive disorder (OCD), post-traumatic stress disorder (PTSD), and/or anxiety as secondary diagnoses. Age range of the sample was 14-45 years, with six females and four males. All subjects had undergone a 5-day intensive course of functional rehabilitation consisting of qEEG directed neurofeedback (NFB), Interactive Metronome (IM) training, and specific neurological rehabilitation (NR) consisting of oculomotor, balance/vestibular, complex motor, core/postural, and dual-tasking therapies (physical activities combined with cognitive exercises). Pre and post cognitive batteries were administered with approximately six weeks duration between tests. Key performance indicators of executive function (EF), cognitive flexibility (CF), simple and complex attention (SA/CA), processing speed (PS), and reaction time (RT) were analyzed. Nutritional support was provided for the duration of intensive training to support metabolic capacities.

Results: After just five days of intensive neurological training, averaging approximately fifteen hours per subject based on three hourly sessions per day, gains in cognitive proficiency were demonstrated at levels ranging from approximately 10-60+%. To eliminate the probability of learning bias, post-testing was performed no earlier than six weeks from the date of pre-test.

Conclusion: Balance, eye movements, motor timing, complex motor skills, core/postural stability, and brainwave activity are inextricably linked to higher cognitive abilities. Quantification of improvement in key performance indicators of cognition in PCS is directly related to objective gains in the above-mentioned areas – demonstrating the efficacy of functional neurological approaches towards conditions that are largely underserved from a conventional medical standpoint. Further study in this area is needed and strongly indicated by these promising outcomes.

In: Movement 2019: Brain, Body and Cognition
Editors: Joav Merrick and Gerry Leisman
ISBN: 978-1-53616-714-6

Chapter 138

FUNCTIONAL NEUROLOGICAL REHABILITATION FOR VESTIBULAR DISORDERS

Omer Hirsh*
Private Practice, Holon, Israel

Keywords: vestibular disorders, function neurology, rehabilitation, primitive reflexes

Vestibular disorders can be detrimental to our well-being and functionality and have long term effects. Vestibular system dysfunction can be due to mechanical causes, such as in benign paroxysmal positional vertigo, infectious causes such as in labyrinthitis, autoimmune causes such as in Meniere's disease, traumatic causes such as a traumatic brain injury, vestibular strokes and more. They can also be secondary to other conditions, such as cerebellar or brainstem strokes, eye movement dysfunction, developmental in origin, such as seen in ADHD, dyslexia, autism and other conditions.

Rehabilitation of the vestibular system must include several components: 1) Removal of the noxious stimulus, if there is one (mechanical/autoimmune/infectious etc), or correcting developmental delays such as residual primitive reflexes, 2) controlling peripheral and central inflammatory processes, 3) improving function of affected components and 4) reestablishing proper integration throughout the vestibular system and its projections to the central nervous system.

In a functional neurological setting, the first part will include repositioning maneuvers, antibiotics treatment when necessary, removal of relevant chemical irritants or foods to which there is sensitivity accordingly, and the examination of retained primitive reflexes etc.

Part two will include nutritional support to reduce inflammation, to promote immune system function and balance, provide substrate for neuronal health and rehabilitation etc. Other tools may include Low Light Laser Therapy, meditation and breathing and more.

Part three will include interventions aimed at selectively activating the affected aspect of the vestibular system, be it a specific canal, the otolithic organ or the vestibulo-cerebellum. This includes rotation or translation in specific directions and speeds with the head neutral or in specific positions so as to activate one canal or one aspect of the otolithic organ over the rest. This can be supported, augmented and enhanced by using appropriate visual stimuli that have reflexogenic relationships with the vestibular system, such as eye tracking, saccades, optokinetic stimulus, visual fixation and more.

In order to ensure long-term effect, integration is important to achieve, by including multiple systems in the rehabilitation process, including the entire cerebellum, as well as the frontal and prefrontal cortex, once it is metabolically and neurologically possible. Implications for the integration of movement and cognition to vestibular function will be discussed.

In: Movement 2019: Brain, Body and Cognition
Editors: Joav Merrick and Gerry Leisman
ISBN: 978-1-53616-714-6

Chapter 139

THE IMPACT OF DIGITAL THERAPY ON MOVEMENT AND COGNITION

Kyle Daigle*
SNA Technologies, Austin, Texas, United States of America

Keywords: functional neurology, digital therapy, movement, video games

Objective: To demonstrate the effectiveness of playing video games using specific light and sound to increase brain function in individuals with neurological deficiencies.

Methods: Playing video games has become a very popular form of entertainment. It is simple to learn and can be very rewarding to the individual playing the game. Light and sound stimulate the brain and different frequencies can have different effects. For example, alpha range visual and auditory stimulation has been shown to reduce the perception of pain. Light and sound have been used to produce a therapeutic benefit for years. The use of different frequencies has been shown time and time again to induce a parasympathetic state. This is beneficial for a range of disease processes from chronic pain, traumatic brain injuries, to decreasing the effects of involuntary movements and neurodegeneration.

Can a video game designed with specific frequencies of light and sound have a targeted effect on functions like balance, coordination and cognition? We know that certain frequencies of light and sound produce different therapeutic effects on the nervous system, and we know that video games can produce positive change in brain function. Combining video games with light and sound therapy specific to certain conditions such as improving balance, coordination, cognition, and movement is what this method will discuss.

Results: There is research that shows many positive effects gaming has on different areas in the brain. It has been shown to activate areas in the brain that deal with semantic memory, visual imagery, and cognitive control. Gaming has also been shown to activate areas related to attention, executive function, and memory. In addition to this, gaming has been found to increase gray matter in the hippocampus. One study was designed to look at balance and the effect video game training for balance had in a group of children with Autism. The results showed that the video game improved balance in the group being studied.

Conclusion: Playing video games has a widespread effect on the brain and creates neuroplastic change in different areas. According to the Life University NeuroLife Institute, playing specific video games using light and sound is not only safe to use but effective in increasing brain function, such as cognitive and balance parameters, in said individuals.

In: Movement 2019: Brain, Body and Cognition
Editors: Joav Merrick and Gerry Leisman
ISBN: 978-1-53616-714-6

Chapter 140

THE USE OF EARLY MOVEMENT VOCABULARY FOR REHABILITATION

Trude Cone*
Amsterdam University of the Arts, Hochschule für Musik und Darstellende Kunst, Frankfurt am Main, Germany

Keywords: early movement progression, rehabilitation, transformational

Objective: I work with clients with functional disconnect problems, such as balance issues, not being able to sequence events, following instructions. These challenges manifest in their ability to be part of society. I will highlight one case study as an example. Since 2017 I have been traveling once a month to work with a 43 year old Greek woman living in Athens, confined the past 17 years to a wheelchair since being hit by a car on her bike in Amsterdam. She was about to graduate as a dancer/choreographer from the School for New Dance Development. She is not paralyzed, but has had surgical fusion of her first two vertebrae which restricts neck motion. A crucial place for regulation of early movement and reflex patterns. During the 17 years she had been treated at many rehabilitation centers in Europe but has only regained a minimal ability to participate, or walk independently. This also has affected her ability to communicate. The rehabilitation centers taught her to propel herself forward, but would never be able to balance without support from others. A way of walking you do not wish on any dancer for the rest of their lives. At the time we started she was hardly able to respond or move independently and was confined to a wheelchair. In my opinion the lack of success was because the therapy focused on the end goal of walking and/or separate components of walking instead of re-engaging the complex movement sequence from the ground up. This is the way we originally learned to walk. Since 2017 we have worked monthly for three days, three hours each day in a movement studio. In the weeks that I am not there I trained a teacher to work with her twice a week for 2 hours.

Result: At this moment she has regained the ability to independently move to the ground from a chair, and use the movement sequences to rise into a side sit and stand on knees. She is able to stand on her feet and balance for 5-10 minutes. As she has gained more independent movement, she also began to participate and communicate better. Before she couldn't move and described being stuck in an in between place. When she reactivated homologous movements her speech began to improve. Over time she began to follow and participate in conversations better, and feel more connected to her surroundings. Also important to note is as her movement sequences began to link together better, as to form a dance, her sentences had more rhythm and she was able to add to conversations in a more coherent way. She is less self-conscious and her attention span and access to vocabulary has improved.

Conclusion: Over the past two years it is evident that re-establishing the early movement sequences enables not only walking, but also renews participation, communication and cognition. Psychological and behavioral factors began to progress again that were also stuck for a long time, as she regained her mobility and voice.

In: Movement 2019: Brain, Body and Cognition
Editors: Joav Merrick and Gerry Leisman
ISBN: 978-1-53616-714-6

Chapter 141

Migraine Treatment with Eurythmy Therapy

Katharina Gerlach** and Urs Pohlman

Institute for Integrative Medicine, Faculty of Medicine, University of Witten/Herdecke, Herdecke, Germany and Institute for Eurythmy Therapy, Faculty of Art-Therapies, Alanus University of Arts and Social Sciences, Alfter, Germany

Keywords: migraine, migraine treatment, eurythmy-therapy, movement-therapy, mind-body-therapy

Objective: The third most common disease in the world is migraine with a global prevalence estimated of 14.7% and more prevalent than asthma, diabetes, and epilepsy combined. Conventional drug treatments can reduce the frequency and severity of attacks, however, with a limited success rate. The satisfaction rate of migraine patients with their conventional treatment is below 50%. Eurythmy therapy (EYT) is a self-activating holistic movement orientated mind-body-therapy that offers therapeutic support for a wide range of symptoms over the past 90 years and is generally applied over a period of 10-12 weeks once a week. It is based on eurythmy as a movement-dance-art, founded in 1911 and taught at schools and universities. Since 1922 Eurythmy therapists (ET) work in practices, clinics and schools. Regarding migraine there are common experiences with successful and sustainable treatments with EYT. In an exemplary case study a woman aged 45 suffered from migraine attacks at least twice a month for 10 years. Long- term medication was prescribed for prophylaxis and drugs to be taken when required for treating the attacks, however with little efficacy. The treatment with EYT over 10 weeks led to a lasting improvement, with a significant reduction in drugs for treating the attacks, persisting at follow-ups at 6 and 12 month control. As this was a single case further case studies remain to be collected.

Method: Over a period of five months, data were collected from 10 different ETs on single-case treatments with EYT on various symptoms: applied exercises, development in movement qualities and patient reported quality of life using a validated questionnaire.

Results: 9% of the patients in this pilot study suffered from migraine. Their data show improvement in movement qualities and also improvement in symptom development and self-perceived quality of life. Applied exercises covered a range of vitalizing, stabilizing, structuring and relaxing effects.

Conclusion: The data from this pilot study's clinical evidence for treatment success with EYT shows its potential in this exemplary field and provides a promising showcase for a relevant understanding of disease and treatment concepts. Migraine is a common disease without a satisfactory treatment from conventional medicine. Thus, an extension of treatment concepts would shed new light on EYT.

In: Movement 2019: Brain, Body and Cognition
Editors: Joav Merrick and Gerry Leisman
ISBN: 978-1-53616-714-6

Chapter 142

SOMATIC PILATES: THINKING WITH OUR BODY, MOVING WITH OUR BRAIN, WORKING WITH OUR HEART

Michal Huber-Rotschild*
Movement Practitioner, Educator and Presenter, Wingate Institute, the National Institute for Excellence in Sports, Natanya and Faculty of Art, Tel- Aviv University, Tel-Aviv, Israel

Keywords: mindful movement, embodiment, neuroplasticity, cognition, Pilates

Objective: In the last 20 years Pilates became a very popular body-mind way of practicing all over the world. The method that was invented by Joseph Pilates (1883-1967) in the 20th century, by the name of "Contrology" originally proposed a way of being and practicing in the world. Constantly striving to acquire strong, healthy bodies while developing our minds to the limit of our abilities in order to achieve the highest accomplishments within the scope of our capabilities in all walks of life- was the message. It has become common practice to teach Pilates as being founded on six fundamental principles: Breathing, centering, concentrating, control, flow and precision.

Methods: Using somatic tools and awareness to all six principles of Pilates, while practicing and teaching Pilates to enable an embodiment process. Paying attention to the body and the movement it has to perform in order to make the trainee more alert and enable him to perceive data from his body, to translate it into better movement, to sense, to feel, to vision the accomplishment of the movement and by that to create neuro-motor development.

Results: My intention is to provide movement teachers, trainers and practitioners and dancers, tools to use the Pilates method in a smarter and deeper way as it should, and by that to gain all the advantages that this method has to offer.

Conclusion: Practicing somatic Pilates can help people regain their movement potential after injury, or rehabilitation can maximize the full human movement capabilities with the minimum effort needed, can improve body awareness and enhance the human nervous system.

In: Movement 2019: Brain, Body and Cognition
Editors: Joav Merrick and Gerry Leisman ISBN: 978-1-53616-714-6

Chapter 143

ASSOCIATED CHANGES IN PROCEDURAL MEMORY OF PARKINSON'S PATIENTS AFTER PARTICIPATION IN ROCK STEADY BOXING

Christopher K McLeod, Atif Ghaffar, Rosemary Gallagher and Adena Leder***
Department of Osteopathic Medicine, New York Institute of Technology, College of Osteopathic Medicine, Old Westbury, New York and School of Health Professions, Doctor of Physical Therapy Program, New York Institute of Technology, Old Westbury, New York, United States of America

Keywords: boxing, rock steady, procedural memory

Objective: Procedural memory (PM) is important for daily function. For example, when riding a bicycle, or typing. Studies show that people with Parkinson's disease (PD) show deficits in PM, in particular, motor PM. There is a known relationship between exercise and PM. Rock steady boxing (RSB) is a group exercise program geared specifically for people with PD. One case series showed improvements in quality of life, balance and gait after participation in RSB. However, the impact of RSB on memory, particular, PM, has not been investigated. Therefore, this study investigated if participation in RSB improves PM.

Methods: A sample of 28 people with PD was recruited and separated into an experimental group planning to attend RSB and another group not planning on attending RSB. To assess procedural learning, a modified Serial Reaction Time Task script was used. Participants were timed as they identified seven blocks of 10 stimuli with a 30 second break between blocks. Procedural learning was assessed by comparing reduction in response time over the blocks.

Results: A total of 28 participants; experimental group n = 14, controls n = 14, participated in the pilot phase of the study. A two-way ANOVA with repeated measures revealed moderate effect of group over the four learning blocks ($p = .19$) indicating that subjects participating in RSB tended to demonstrate faster reaction. Random sequence control subjects showed on average a 93.5ms decrease in median reaction time, while subjects participating boxing showed an increase in median reaction time of 27.3ms.

Conclusion: Pilot data indicates group exercise programs such as RSB are promising, however further data collection is needed and in progress to determine a positive role in improving PM & learning in patients with PD. The results indicate RSB participants serially performed better with shorter reaction times compared with PD patients not participating in RSB. This is a novel finding. Current literature lacks on PM improvement data in PD patients. Further studies are needed to evaluate SRTT and new modalities to improve SRTT in PD patients.

In: Movement 2019: Brain, Body and Cognition
Editors: Joav Merrick and Gerry Leisman
ISBN: 978-1-53616-714-6

Chapter 144

WHY THE BODY'S OLDEST SYSTEMS ARE THE KEY TO SOLVING OUR NEWEST PROBLEMS

Karla Mehlenbacher*
Take Root Brain and Body Development In,
Denver, Colorado, United States of America

Keywords: functional neurology, neuroplasticity, primitive reflexes, hemisphere, behavior, movement, cognition

Objective: This presentation seeks to demonstrate a viable link between movement and cognition. Movement is the language of the brain. Based on this insight, extensive research and clinical neuroscience, there are simple ways to assess, understand, and work with the body's non-verbal messages to improve cognitive functions. Embodied neuroplasticity is the idea that we can repair and renew cognitive and emotional functions with specific brain-based movements.

Methods: A sample of six individuals was chosen from our database with attention deficit-hyperactivity disorder (ADHD) as the primary diagnosis; and anxiety, depression, obsessive- compulsive disorder (OCD), and/or mild cognitive impairment as secondary diagnoses. Age range of the sample was 25-35 years, with three females and three males. All subjects had undergone a two month course of embodied neuroplasticity consisting of primitive reflex integration, activation of the parasympathetic system, Interactive metronome (IM) training, and specific neurological rehabilitation (NR) consisting of complex motor and dual-tasking therapies (physical activities combined with cognitive exercises). Pre and post cognitive batteries were administered with 8 weeks duration between tests. Key performance indicators of executive function (EF), cognitive flexibility (CF), simple and complex attention (SA/CA), processing speed (PS), and reaction time (RT) were analyzed.

Results: Improved executive function (EF), cognitive flexibility (CF), simple and complex attention (SA/CA), processing speed (PS), and reaction time (RT). Subjective decreased anxiety, relaxation, improved focus, decreased muscle tension, improved digestion and bowel movements, decreased heart rate, improved balance and mood.

Conclusion: Primitive reflex integration, motor timing, complex motor skills, and parasympathetic activation are intricately linked to higher cognitive abilities. Quantifiable improvement in key performance indicators of cognition in ADHD are directly related to objective gains in the above mentioned areas demonstrating the efficacy of embodied neuroplasticity. Further study in this area is needed and strongly indicated by these promising outcomes.

In: Movement 2019: Brain, Body and Cognition
Editors: Joav Merrick and Gerry Leisman
ISBN: 978-1-53616-714-6

Chapter 145

OVERCOMING STAGE-FRIGHT THROUGH THE BODY, USING LABAN-BARTENIEFF MOVEMENT SYSTEM

Sharon Gidron Peskin*
Private Practice, Herzliya, Israel

Keywords: stage fright, Laban-Bartenieff Movement System

They say that most people would rather be in a coffin than give a eulogy at the funeral… stage fright has a huge toll on self-esteem and often compromises career or life choices. This workshop will enable the participants to explore movement patterns of both fear and charisma, and to discover paths of shifting from helplessness to confidence through body engagement. When experiencing stage-fright, the body seems to wildly get out of control. It makes you feel exposed and useless. It is quite a rough experience.

In this workshop I will demonstrate how to deconstruct stage-fright, which is a complex phenomenon, into simple body patterns, in order to re-pattern the body reactions when dealing with public speaking.

My goal is to awaken body awareness within the setting of public-speaking, a setting that isn't considered as 'physical'. I believe in a playful learning process, filled with curiosity and grace. Laban-Bartenieff Movement System is a fundamental resource in my work.

Topics/exercises we may experience or discuss:

- Diagnosing personal stage-fright movement patterns through Laban-Bartenieff Movement System.

- Turning space into an ally rather than an enemy – through scales, space effort and shape (inner space).
- Empowering presence through breath support and weight sensing.
- Allowing yourself to really engage through shaping and bridging.
- Unlocking the bound flow and enabling natural and authentic interaction.
- Enabling better inner-outer regulation and control through core-distal work and X-O practice.

In: Movement 2019: Brain, Body and Cognition
Editors: Joav Merrick and Gerry Leisman
ISBN: 978-1-53616-714-6

Chapter 146

SHOULD BALLET DANCERS VARY POSTURES AND UNDER-FOOT SURFACES WHEN PRACTICING POSTURAL BALANCE?

Nili Steinberg*[*] *and Oren Tirosh
The Wingate College of Physical Education and Sport Sciences at the Wingate Institute, Netanya, Israel and School Medical and Health Sciences, Swinburne University of Technology, Australia

Keywords: postural balance, accelerometers, dancers, textured/spiky surfaces, proprioception

Objective: Postural balance (PB) is an important component skill for professional dancers. However, the effects of different types of postures and different underfoot surfaces on PB have not been adequately addressed. The main aim of the present study was to investigate the effect of different conditions of footwear, surfaces, and standing positions on static and dynamic PB ability of young ballet dancers.

Methods: Thirty-six male and female young professional ballet dancers (aged 14-19 years) completed static and dynamic balance testing, measured by head and lumbar accelerometers, while standing on one leg in the turn-out position, under six different conditions: 1) "relaxed" posture; 2) "ballet" posture; 3) barefoot; 4) ballet shoes with textured insoles; 5) barefoot on a textured mat and 6) barefoot on a spiky mat.

Results: A condition effect was found for static and dynamic PB. Static PB was reduced when dancers stood in the "ballet" posture compared with standing in the "relaxed" posture, and when standing on a textured mat and on a spiky mat ($p < .05$); and, Static PB in the "relaxed" posture was significantly better than PB in all the other five conditions tested. Dynamic PB was significantly better while standing in ballet shoes with textured insoles and when standing on a spiky mat, compared with all other conditions ($p < .05$).

Conclusion: The practical implications derived from the present study are that both male and female dancers should try to be "relaxed" in their postural muscles when practicing a "ballet" aligned position; including dance practice on different types of floors and on different types of textured/spiky materials may result in skill transfer to practice on normal floor surfaces; and, both static and dynamic PB exercises should be assessed and generalized into practical dance routines.

In: Movement 2019: Brain, Body and Cognition
Editors: Joav Merrick and Gerry Leisman
ISBN: 978-1-53616-714-6

Chapter 147

The Oriental Concept of Smooth and Harmonious Qi Flow in the Meridians for the Optimization of the Elite Athlete's Movement and for Sports Injury Prevention

Karl Zippelius*
Sports Medicine, Praxis Florenz, Firenze, Italy

Keywords: Qi, sport, athlete, movement, injury, prevention

Western approach to movement and locomotor function is mainly based on biomechanics, molecular/metabolic processes and signal transmission of the nervous system. Any system has its limits to explain reality. Applying the aforementioned, western' models it is difficult to explain why it seemed to be helpful to Tour de France winner Vincenzo Nibali to undergo acupuncture every day to increase performance and reduce pain or why US swimmer Michael Phelps insisted to get Chinese cupping therapy during Olympic Games in Rio. Recently Austrian professional soccer teams started to insert yoga exercises into athlete preparation to reduce risk of injuries and optimise movement with good results, but without scientific explanation of the positive effects.

Chinese medicine with its idea of 'Qi' flow through the body and basic concepts like meridian network, yin/yang polarity and five elements model, offers a further and different key to understand physiology of body and how to explain positive effects of fore mentioned techniques on movement and athletes health.

The aim of this workshop is to offer another point of view on how to optimize elite athlete's movement and prevent sports injuries. Basic concepts of TCM with reference to locomotor system (e.g., meridian diagnosis, Qi flow) and movement (e.g., Tai Qi, Qi Gong) will be explained and illustrated by case reports of TCM athletic preparation and treatment of sports professionals by acupuncture, external application of Chinese herbs, hand techniques and energetic movement.

In: Movement 2019: Brain, Body and Cognition
Editors: Joav Merrick and Gerry Leisman
ISBN: 978-1-53616-714-6

Chapter 148

THE IMPACT OF AEROBIC AND NON-AEROBIC WALKING ON COGNITIVE ABILITIES IN SCIENCE IN ELEMENTARY SCHOOL CHILDREN

Raed Mualem[1,2,*], Shiren Rustom[1] and Gerry Leisman[2-4]
[1]Oranim Academic College, Qirit Tivon, Israel,
[2]Institute for Brian and Rehabilitation Sciences, Nazareth, Israel
[3]Department of Physiotherapy Faculty of Social Welfare, University of Haifa, Israel
[4]University of the Medical Sciences, Havana, Cuba

Keywords: effortless movement, movement with effort, brain, cognitive ability

Background: Studies have shown that exercise is a paramount factor in maintaining physical and mental health. Moreover, exercise is a promising strategy for improving cognitive functions and can affect brain plasticity. Recent studies have indicated the significant effects that walking exerts on cognitive function. These studies have indicated that movement in general, and ten minutes of aerobic exercise in particular, has a significant effect in improving memory capacity and attention span.

Objective: Against this backdrop, the purpose of this study was to examine the effect that aerobic and non-aerobic walking has, with and without effort, on memory and attention span as well as on thinking skills in science subjects.

Methods: The study included thirty, fourth grade children in an elementary school in the Arab sector in the Haifa district in Israel. The study employed the Simon game to test memory, a standardized hidden figures feature detection task for examining attentional focus and the three questionnaires that test the ability to think at five Bloom Taxonomy levels of: knowledge, understanding, application, analysis and synthesis.

Results: The results of the study indicate that effortless walking significantly increases memory and attentional focus as measured by standardized feature detection tasks by 2.5 and 1.2 times, while walking with effort resulted in attentional focus and memory values not significantly different to those obtained at rest. However, when examining high-order skills: effortless walking demonstrated a significant increase in synthesis and analysis skills by 1.6 and 2.2 respectively, while walking with effort demonstrated no significant differences to the values obtained at rest.

Conclusion: The results indicate that simple ten-minute walks prior to school-based learning tasks will have a significant effect on improving cognitive performance in elementary school-aged children. The study along with others justifies the importance of physical education within the school system.

In: Movement 2019: Brain, Body and Cognition
Editors: Joav Merrick and Gerry Leisman
ISBN: 978-1-53616-714-6

Chapter 149

DETERMINING THE NEUROCOGNITIVE STATUS AND THE FUNCTIONAL ABILITY OF PATIENTS TO SCREEN FOR HIV-ASSOCIATED NEUROCOGNITIVE DISORDER (HAND)

Ritika Agarwal*[*]*, Ravinder Singh Aujla*[†]*,
Amit Gupta*[‡] *and Mukesh Kumar
Department of Medicine, Chandra Laxmi Hospital, Vaishali, Ghaziabad

Keywords: patient living with HIV, HIV associated neurocognitive disorder, neurocognitive assessment, dementia

Background: To adequately evaluate the extent of neurocognitive impairment, a battery of neuropsychological tests is typically administered which are neither cost effective nor time efficient in the outpatient clinical setting. The aim of the study was to assess neurocognitive status and functional ability of people living with HIV and find a brief screening tool to identify those who would benefit from a full diagnostic evaluation.

Method: The study enrolled 160 Patient living with HIV (PLHIV) (80 pre-ART and 80 On-ART) fulfilling the inclusion and exclusion criteria. Neurocognitive assessment and an assessment of Functional ability was done by using the Montreal Cognitive Assessment (MoCA) and Lawton and Brody IADL Scale respectively.

Results: The study population consisted of 75.6% males and 24.4% females with mean age of 44 ± 10 years. All of the subjects were literate (61.2% subjects had received

[*] Corresponding Author's Email: Aritika1994@gmail.com.
[†] Corresponding Author's Email: raujla@live.ca.
[‡] Corresponding Author's Email: amitgupta.clh@gmail.com.

up to High school level education), 76.2% were married. Only 5% of subjects were found to have mild impairment in ADL with the most common affected domain being food preparation (50%). The overall prevalence of HAND in the study subjects was 52.5%. Of these, 47.5% had ANI and 5% had MND. In MoCA, the most frequently affected domains were Language (97.6%), visuospatial ability (92.9%) and memory (71.4%).

Conclusion: The prevalence of HAND is one in two PLHIV. The prevalence of HAND among Pre-ART subjects (i.e., ART naive) and those on ART is similar suggesting that neurocognitive impairment starts early in HIV infection. Older age (>40 years) and fewer years of formal education were important determinants statistically associated with the presence of HAND. HAND involves both cortical and subcortical regions of the brain with sub cortical involvement being predominant. Memory and Visuospatial function impairment had the most predictive potential for detecting the presence of HAND. HAND screening is recommended in all PLHIV at enrolment into care. Simple tools like MoCA can be used in busy outpatient settings by healthcare workers to screen for HAND.

In: Movement 2019: Brain, Body and Cognition
Editors: Joav Merrick and Gerry Leisman
ISBN: 978-1-53616-714-6

Chapter 150

TRANSFERRING THE LEARNING OF MOVEMENT OR LEARN ONE THING AND KNOW MANY THINGS

Moti Nativ*
Independent Feldenkrais and Martial Arts Practice, Tel Aviv, Israel

Keywords: movement, learning, Feldenkrais, self-preservation

Objective: This workshop explores how to enhance the quality of our movement and discover alternative ways of using our body through Awareness through Movement (ATM) lessons, developed by Moshe Feldenkrais and experience the ability to transfer the skills learned by applying them in different positions and situations. The genius of Moshe Feldenkrais' work was to translate abstract concepts into physical movements and lived experience. The concept of transferring the quality of a learned movement, common to both the Feldenkrais Method and Judo, suggests a way to create a layer for quicker learning. Knowing that Judo had significant influence on the Feldenkrais Method, we'll use the learned movement to respond to changes in the environment thus assuring our self-preservation. We'll sense the elements of good movement by controlling the timing, speed, direction, and distance.

Method: I will lead the participants through an ATM lesson and discuss fundamental concepts of connecting to the environment through orientation (time and space) and continuity of action (balance and dynamic stability). The movement lesson will focus on improving the function of reaching with the arms and how additional body parts – (shoulders, chest and back, pelvis feet) support this intention. The basic movements will be learned in different positions, such as lying and standing. To ascertain that the learning process was successful and the newly acquired skills will be applied in different positions and situations, emphasizing the connection to the changing environment.

Results: At the conclusion of the process, participants will have acquired a skill that, through the concept of transferring the learning, can be used with confidence in different environments and intention.

Conclusion: The time it takes to learn a new movement can be shortened considerably by using awareness through movement and employing the concept of transferring learning to different situations.

In: Movement 2019: Brain, Body and Cognition
Editors: Joav Merrick and Gerry Leisman
ISBN: 978-1-53616-714-6

Chapter 151

RETAINED PRIMITIVE REFLEXES AND ARTICULATION DISORDERS IN POLISH PRESCHOOL CHILDREN: A PILOT STUDY

Paulina Krzeszewska*
Department of Philology, University of Silesia, Katowice, Poland

Keywords: primitive reflexes, retained reflexes, speech development, articulation disorders

Objective: To describe the co-occurrence of chosen retained reflexes - symmetrical tonic neck reflex (STNR), asymmetrical tonic neck reflex (ATNR), tonic labyrinthine reflex (TLR) - and articulation disorders among preschoolers.

Method: Participants of the pilot study were recruited through two kindergartens in Zabrze. Among the 60 preschoolers aged 5-6 years, 30 did not present articulation disorders, while 30 did. Spoken language production, orofacial anatomy, phonemic awareness and chosen primitive reflexes - STNR, ATNR, TLR - were assessed.

Result: Each child manifesting articulation disorders also presented at least one of three assessed retained primitive reflexes. However, five out of 30 children without articulation disorders presented retained reflexes as well. What is essential, the results show significant co-occurrence, which should not be perceived as cause and effect relationship.

Conclusion: Children with articulation disorders manifest preserved ATNR, STNR and TLR. Nevertheless, presented topic requires a greater research sample to achieve higher reliability and validity of the results. It may become effective to include retained reflexes therapy strategies in speech-language therapy, which also should be verified by empirical research.

In: Movement 2019: Brain, Body and Cognition
Editors: Joav Merrick and Gerry Leisman
ISBN: 978-1-53616-714-6

Chapter 152

INFLUENCE OF PESTEL FACTORS IN THE APPEARANCE OF MUSCULOSKELETAL DISORDERS IN THE MANUFACTURING INDUSTRY: DOCUMENTARY REVIEW

Juan Zamora Medina**, Monica Rodriguez and Claudia Borré
Specialization in Ergonomics of the Colombian School of Rehabilitation Institution,
Santa Fe de Bogota, Columbia

Keywords: musculoskeletal disorders, ergonomics, biomechanical irrigation, manufacturing industry, Occupational Disease.

Objective: To identify, from a documentary review, the influence of the PESTEL factors on the appearance of musculoskeletal disorders in manufacturing industries in order to address this problem from an ergonomic perspective.

Methods: A documentary review focused on the identification of the influence of PESTEL factors on the appearance of EMD in the manufacturing industry. Initially they were used as search words: PESTEL factors, ergonomics, manufacturing industry, environmental factors, musculoskeletal disorders and musculoskeletal disorders. The search was performed in different databases such as Medline, Science direct and open edition, articles were found in English, French and Spanish with publication date between 1989 and 2015.

Results: The articles found during the documentary review were focused on the manufacturing industry, however no articles were found that relate the three fundamental aspects of the documentary review: PESTEL, manufacturing and musculoskeletal disorders. Through the search carried out regarding the influence of the PESTEL factors in the appearance of EMD, we were able to show that there is little bibliographical reference referring to the subject in general and only some studies in which small mentions are made in an individual way, such factors as influential in the generation of DME in the manufacturing industry. From the first review carried out, it was found that, although there was no relationship between PESTEL factors and the appearance of musculoskeletal disorders in the manufacturing industry, it was found in the articles that expose the importance of analyzing factors other than biomechanics, which they are called external factors that are mainly related to psychosocial and organizational factors. At present we have many methods that specifically describe each of the biomechanical factors that intervene in the actions carried out by workers during the performance of their tasks, but we lack scientific studies that help us to show that there really are other factors such as the PESTELs that intervene in the appearance of DME.

Conclusion: We identified a lack of scientific evidence that addresses, from ergonomics, the PESTEL factors to generate intervention models at a systemic level.

In: Movement 2019: Brain, Body and Cognition
Editors: Joav Merrick and Gerry Leisman
ISBN: 978-1-53616-714-6

Chapter 153

MANUAL DIRECTIONAL GESTURES FACILITATE CROSS-MODAL PERCEPTUAL LEARNING

Anna Zhen[1-4], Stephen Van Hedger[4], Shannon Heald[4], Susan Goldin-Meadow[4] and Xing Tian[1-3,*]

[1]Division of Arts and Sciences, New York University Shanghai, Shanghai, China
[2]Shanghai Key Laboratory of Brain Functional Genomics (Ministry of Education), School of Psychology and Cognitive Science, East China Normal University, Shanghai, China
[3]NYU-ECNU Institute of Brain and Cognitive Science at NYU Shanghai, Shanghai, China
[4]Department of Psychology, University of Chicago, Chicago, Illinois, United States of America

Keywords: sensorimotor integration, multisensory integration, lexical tones, gesture, speech learning

Objective: Action and perception interact in complex ways to shape how we learn. In the context of language acquisition, for example, hand gestures can facilitate learning novel sound-to-meaning mappings that are critical to successfully understanding a second language. However, the mechanisms by which motor and visual information influence auditory learning are still unclear. We hypothesize that the extent to which cross-modal learning occurs is directly related to the common representational format of perceptual features across motor, visual, and auditory domains (i.e., the extent to which changes in one domain trigger similar changes in another). Furthermore, to the extent that information across modalities can be mapped onto a common representation, training in one domain may lead to learning in another domain.

Methods: To test this hypothesis, we taught native English speakers Mandarin tones using directional manual gestures. Two types of gestures were included: 1) gestures that were congruent with pitch direction (congruent gesture, e.g., an up gesture moving up, and a down gesture moving down, in the vertical plane), and 2) gestures that were rotated 90 degrees from the congruent gestures (rotated gesture, e.g., an up gesture moving away from the body, and a down gesture moving toward the body, in the horizontal plane). One of four groups (18 participants in each group) either performed or watched one of the two types of gestures.

Results: Watching or performing the congruent gestures significantly enhanced tone category learning, compared to auditory-only training. Moreover, when gestures were rotated (e.g., an up gesture moving away from the body, and a down gesture moving toward the body, in the horizontal plane), performing the gestures resulted in significantly better learning, compared to watching the rotated gestures.

Conclusion: Our results suggest that when a common representational mapping can be established between motor and sensory modalities, auditory perceptual learning is likely to be enhanced.

In: Movement 2019: Brain, Body and Cognition
Editors: Joav Merrick and Gerry Leisman
ISBN: 978-1-53616-714-6

Chapter 154

MOTOR-COGNITIVE DUAL-TASKING: INSIGHTS FROM CEREBELLAR STROKE PATIENTS

Erika CS Künstler**, Carsten M Klingner, Albrecht Günther, Kathrin Finke, Otto W Witte and Peter Bublak
Department of Neurology, Jena University Hospital, Jena, Germany

Keywords: motor-cognitive dual-tasking, visual attention, cerebellum, stroke

Objective: Whilst the cerebellum has long been known to coordinate movements and regulate posture, studies have increasingly implicated the cerebellum in executive functions such as dual-tasking. However, whilst the cerebellum is active during dual-tasking involving the simultaneous execution of a motor and a cognitive task, it is still not entirely clear which precise mechanisms are supported by the cerebellum, although it is thought that the cerebellum may integrate individual task networks into one distinct dual-task network.

Methods: A dual-tasking paradigm was created using a continuous finger-tapping task in addition to a cognitive visual attention task. 26 cerebellar stroke patients and 26 matched controls completed this paradigm, and the dual-task related decline in the both tasks was examined. Effects on visual attention capacity were analysed by estimating participants' visual processing rate and the storage capacity of the visual short-term memory. Furthermore, resting state fMRI data was also collected from 19 of the patients, and 19 of the controls. A test of motor dexterity ensured that patients had no residual motor deficits.

Results: Both groups performed similarly in the TVA-based task in both the single and dual-task conditions, with no significant differences between the groups in their visual processing rate and the visual short-term memory storage capacity. However, despite showing no residual deficits in fine motor skills and performing equally well in the single-task condition, an ANCOVA indicated that patients had significantly more difficulties in performing the motor task in the dual-task condition ($p = 0.02$; $d = 0.68$). A resting state fMRI analysis indicated a correlation between the performance costs in the motor task in the dual-tasking condition and the connectivity between the cerebellum and the ventral attention network in healthy controls ($p = 0.01$, $r = 0.56$, $R2 = 0.31$), but not in patients ($p = 0.20$, $r = -0.31$, $R2 = 0.10$).

Conclusion: Even though both the motor and the visual attention system function as normal under single-task conditions, patients struggle with the additional cognitive load of dual-tasking. Via its connection to attentional networks, the cerebellum plays a decisive role in coordinating cognitive resources across tasks during motor-cognitive dual-tasking.

In: Movement 2019: Brain, Body and Cognition
Editors: Joav Merrick and Gerry Leisman
ISBN: 978-1-53616-714-6

Chapter 155

BODY COGNITION METHOD (BCM): THE IMPROVEMENT OF QUALITY OF LIFE AND COGNITIVE FUNCTIONS, SUPPORTED BY TWO RECENT STUDIES

Yosefa Michaeli*
Founder of Body Cognition Method (BCM), Tel Aviv, Israel

Keywords: cognition, brain, awareness, quality, learning

Objective: Explaining several aspects of BCM and demonstrating two recent clinical trials that support the BCM's premise that, any person practicing the BCM consistently, stands to gain a significant cognitive advantage as well as a better quality of life over those who do not. Once a person learns the correct way of movement, he develops an acute properioseptic sense, enabling him to achieve high awareness of his physical behavior, resulting in exceptional physical and mental performance. This system provides the sensory feed which enables us to orient ourselves in space in every aspect: distinguishing between the relations of our body parts to each other, to the space around us while targeting our movements and grasping the gravity line. This complex system is crucial to our physical conduct.

Method: Two clinical trial studies.

Result: One study compared adults practicing BCM exercise to those practicing other exercise. As predicted, both participants showed equal levels of general quality of life while only the participants engaged in BCM exercise showed higher cognitive functions such as phonetic fluency, backward digit span and other tasks. The other study also compared adults practicing BCM exercise to adults practicing other form of exercise. The results of this study show that BCM followers displayed better functions including higher ratings of life quality and higher cognitive functions such as, attention span, language skills, memory and managerial functions.

Conclusion: With nearly 60 years of experience and field work, we have seen thousands of pupils consistently practice the BCM so that we can clearly examine their state of health with the passage of time. Consequently, we are confident in stating that these individuals remain lively, continue to work, rarely undergo orthopedic operations despite their advanced age and are much less accident prone. In our experience, the continuous participation in BCM lessons has proven benefits in preventing health problems, maintenance of physical and mental capacities and when required, rehabilitation.

In: Movement 2019: Brain, Body and Cognition
Editors: Joav Merrick and Gerry Leisman
ISBN: 978-1-53616-714-6

Chapter 156

RHYTHMIC LIMB MOVEMENTS AFFECT SPEECH OUTPUT DIFFERENTLY IN INDIVIDUALS

Heather Weston[1,2,*] and Susanne Fuchs[1]
[1]Leibniz-Centre General Linguistics (ZAS), Berlin
[2]Humboldt-Universität zu Berlin, Germany

Keywords: speech production, hand–speech link, embodied cognition, individual differences

Objective: Various studies have reported links between physical activity and enhanced cognitive performance in certain areas. Yet fewer studies have examined whether the cognitive process of speech production is affected by contexts of movement and, if so, at which linguistic levels. This pilot study starts by investigating the level of speech sounds, asking whether self-paced cycling affects how speech is produced across time: specifically, does movement affect speech rate, the duration of speech chunks and pauses, and the amount of speech produced?

Methods: Acoustic, respiratory and motion capture data were collected from 15 (7 male) adult speakers of German in a laboratory setting. An item-choice task was used to elicit spontaneous speech in three conditions: 1) speaking only; 2) hand-pedaling while speaking; and 3) foot-pedaling while speaking. The use of two different movement conditions was motivated by prior research suggesting a hand–mouth link in spoken speech. The acoustic data (˜ 2 min./trial; 9 trials/participant) were transcribed and segmented into speech chunks and pauses (>130ms) using phonetic analysis software. Words were segmented into syllables and speech rate (syllables/second) was calculated for each speech chunk.

Results: Speech rate did not increase with movement but was similar across conditions, as was pause duration, though there was less variability during movement. Movement was, however, associated with an increase in the number of words and duration of speech chunks, though individual speakers behaved differently, particularly in terms of number of words produced: 5 speakers showed no change between conditions; 4 speakers produced more words during both movement conditions and 3 speakers in one movement condition; 2 speakers produced fewer words during hand-pedaling.

Conclusion: The findings suggest that movement encourages online production of speech for some speakers. In an embodied perspective, body movement may also keep speech "moving". But it is not clear why individuals reacted differently to the movement conditions. One possibility is that idiosyncratic physical properties play a role, e.g., by increasing cognitive load (dual task) for participants with lower fitness or less cycling experience. Future work will investigate syntactic complexity and use different activity tasks and effort levels.

In: Movement 2019: Brain, Body and Cognition
Editors: Joav Merrick and Gerry Leisman
ISBN: 978-1-53616-714-6

Chapter 157

SHARED MOVEMENT APPROACH: DANCE/MOVEMENT INTERVENTIONS INTO INTERPERSONAL ATTUNEMENT WITH CHILDREN AND ADOLESCENTS ON THE AUTISTIC SPECTRUM

Rosemarie Samaritter*
Codarts University of the Arts, Rotterdam Netherlands

Keywords: movement analysis, interpersonal attunement, intersubjectivity, autism, dance, movement therapy

Background: Interpersonal relating in autism is characterised by atypical attunement patterns, which affect the self-other-interactions and the development of intersubjectivity. In dance movement therapy (DMT) dynamic and rhythmic movement activities are used to support the integration of body-mind and to address the nonverbal interpersonal relationship between participants.

Objective: Outcome of retrospective movement analysis of interpersonal relating in ASD participants during DMT intervention was used to define movement markers that convey interpersonal relatedness and to design a DMT intervention model that addresses interpersonal engagement in participants with ASD.

Methods: Retrospect open coding with Laban based movement analysis was used to collect specific qualities of interpersonal relating behaviours of ASD participants during dance movement therapy. In a second study a set of selected movement markers were applied in the observation of interpersonal relating behaviours in ASD participants in a series of N = 1 modelled DMT intervention. For pre-post treatment evaluation the outcome of movement observation was triangulated with the outcome of standardised social behaviour questionnaires.

Results: The retrospective analysis of DMT with ASD's yielded a set of movement markers/measures for the changes in interpersonal/social engagement and attunement behaviours (SEAM), which were summarized in the SEAM observation scale. This scale yielded sufficient to good interrater agreement and reliability during interval coding procedures.

The outcome of DMT intervention with regard to the development of interpersonal relating behaviours was replicated in a series of single cases. SEAM markers have shown suitable to monitor changes of interpersonal relating behaviours initiated by the ASD participant.

Conclusion: Movement markers for social engagement and attunement behaviours may enable therapists and caregivers to monitor interpersonal relating behaviours from a movement perspective. Compared to the usual social cognition models, this action-oriented model brings an inactive perspective to interpersonal engagement and attunement and can therefore be applied directly during the interaction with the ASD child. The results form clinical outcome monitoring and a pilot study in a series of single cases show encouraging improvements in interpersonal engagement and non-verbal attunement behaviours in participants with ASD.

In: Movement 2019: Brain, Body and Cognition
Editors: Joav Merrick and Gerry Leisman
ISBN: 978-1-53616-714-6

Chapter 158

EFFECTS OF A MYOFASCIAL SELF-MASSAGE OF THE PLANTAR FASCIA ON THE MOBILITY OF THE HIPS DEPENDING ON DIFFERENT MASSAGE DURATIONS

Tessa Temme**, Denise Temme and Christian Büning
German Sports University, Cologne, Germany

Keywords: fascial network, flexibility, range of motion, plantar fascia

Objective: Research on the subject of myofascial training is raising. As part of this, myofascial self-massage (SMR), in which a pressure stimulus is exerted on the fascia tissue by foam roller or a ball, is suspected to enhanced mobility. The acute and long-term effects on mobility and strength abilities have yet not been sufficiently clarified. Some studies examine the extent to which massage influences on plantar fascia affect more distant joints due to the network of myofascial chains. It has been discovered that a two-sided foot massage causes an immediate and significant increase in the range of motion (ROM) of the hip joints as well as the lower back. Further studies examined the influence of a one-sided foot massage on the dorsiflexion of the ankle joint. However, no significant results could be found. In the aforementioned studies, the subject numbers were rather small (around 30). The purpose of this study is to investigate the effects of myofascial-self massages in greater detail. To do so, we collected a large sample and investigated the effects of different massage durations. We are raising the question if changes in duration also lead to changes in the remote effects (e.g., ROM of the hip joints)? Furthermore, we investigated if acute effects were still detectable 15 minutes after the intervention?

Methods: Single-blind, randomized, controlled trial with 400 subjects. To evaluate the mobility of the participants, a sit-and-reach (SaR) test was conducted using a Sit-and-reach measuring box. After the mobility test, the two test groups used a lacrosse ball for the self-massage of the plantar fascia while standing (duration: 2min/4min). The control groups were standing for the same time interval. Subsequently, a second SaR-Test after the intervention and a third SaR-Test after 15 minutes were carried out.

Results: A GLM with repeated measures was carried out. While comparisons of the whole Group showed no significant differences between the intervention groups, dividing the group into rather mobile and rather immobile participants, showed significant intervention effects of the different self-massage durations. We conclude that the rather immobile participants benefit vastly from the ball massage, while the rather mobile subgroup does not.

In: Movement 2019: Brain, Body and Cognition
Editors: Joav Merrick and Gerry Leisman
ISBN: 978-1-53616-714-6

Chapter 159

FLUENCY THROUGH SYNCHRONY: A COGNITIVE ACCOUNT OF THE SOCIAL FUNCTIONS OF DANCE

Joshua Bamford[1,*], Bronwyn Tarr[1,2] and Emma Cohen[1]
[1]Institute of Cognitive and Evolutionary Anthropology, University of Oxford, Oxford
[2]Department of Experimental Psychology, University of Oxford, Oxford, United Kingdom

Keywords: dance, evolution, synchrony, social bonding, social cognition

Objective: It has been suggested that one of the primary evolutionary functions of music-dance is that it promotes social bonding and group cohesion, which enhances group survival and individual fitness. Music-dance activity requires a range of cognitive abilities, among them a capacity for sensorimotor synchronisation. Previous research suggests that interpersonal synchronisation may have specific social bonding effects, often described as a sensation of self-other overlap. However, there is an incomplete understanding of why synchronised movement should lead to this subjective experience. The present research outlines a novel theory to explain the synchrony-bonding effect – which behaviourally synchronised interactions may be easier to process than non-synchronised interactions through a process of perceptual streaming, leading to a rewarding sensation of fluency and positive associations with the other.

Methods: In study 1, participants complete a modified flanker task using animated Gabor patches, in which participants must determine the direction of movement. In study 2, in a dual task paradigm, participants perform a finger tapping task with two virtual partners while also completing a visual attention task. The finger tapping task is manipulated through varying the level of synchrony in the interaction, and the resultant effect on performance in the attention task is recorded.

Results: Testing is ongoing - preliminary results will be presented. It is expected that participants in study 1 will have faster reaction times and lower error rates in congruent than incongruent conditions. In study 2, it is expected that participants will perform better on the visual perception task when their tapping partners are in synchrony with each other than when they are not.

Conclusion: Results in line with hypotheses will support our novel account of the multi-modal perception of synchrony, suggesting that experiencing synchrony of movement results in the perceptual streams of self and other becoming aligned. This is then aesthetically pleasing because processing a single stream is less cognitively taxing than multiple streams created by non-synchrony. This rewarding experience may then be associated with other agents in the interaction, leading to a social bonding effect. Ultimately, this may contribute to our understanding of the evolutionary functions of music-dance.

In: Movement 2019: Brain, Body and Cognition
Editors: Joav Merrick and Gerry Leisman
ISBN: 978-1-53616-714-6

Chapter 160

INTERHEMISPHERIC INTERACTIONS ACROSS BASAL GANGLIA AND CEREBELLAR CIRCUITS DURING MOTOR CONTROL

Ana Cristina Vidal[1-3,*], Paula Banca[3], Augusto Gil Pascoal[1], Gustavo Cordeiro[4,5], João Sargento-Freitas[4] and Miguel Castelo-Branco[2]
[1]Universidade de Lisboa, Fac Motricidade Humana, Lisbon, Portugal,
[2]Garcia de Orta Hospital, Almada, Portugal,
[3]Centro Interdisciplinar de Investigação Egas Moniz,
[4]Visual Neuroscience Laboratory, IBILI, Faculty of Medicine, University of Coimbra, Portugal
[5]Stroke Unit, Department of Neurology, Coimbra University Hospital, Coimbra, Portugal

Keywords: sub-cortically modulation, arm elevation, upper limb restraint, physiotherapy, motor control

Understanding the nature of subcortical interactions during motor facilitation and/or restriction may have important functional implications concerning in health and disease. Here we investigate this question by examining how these circuits modulate when the whole upper limb is recruited to perform unilateral arm elevation, in both dominant and non-dominant arms, as a function of presence/absence of restriction of the contralateral limb. We used this paradigm to study the subcortical modulation in 17 healthy subjects using functional magnetic resonance imaging (fMRI). We found the expected activation of contralateral motor areas and ipsilateral cerebellum for unilateral arm elevation. However, only dominant limb elevation leads to contralateral striatal deactivation. In presence of contralateral upper limb restraint we found activation of the caudate nucleus. Only elevation of the dominant arm elicited deactivation of the subthalamic nucleus, irrespective of the presence or absence of restraint. In contrast, non-dominant limb elevation only elicited striatal deactivation during contralateral restriction. Contralateral cerebellar activation was only present for the non-dominant limb during restraint while deactivation was observed for the dominant limb.

In sum, we found a striking subcortical activation/deactivation dependence on hemispheric motor dominance. The addition of contralateral limb restriction led to a profound modification of subcortical recruitment patterns. Thus, subcortical patterns can be modulated by motor facilitation or restraint procedures as a function of hemispheric dominance. These novel physiological findings are relevant to understand motor control and potentially important for clinical decision making in neurorehabilitation.

In: Movement 2019: Brain, Body and Cognition
Editors: Joav Merrick and Gerry Leisman
ISBN: 978-1-53616-714-6

Chapter 161

CONCEPTS OF POSITIVE AND NEGATIVE NEUROPLASTICITY AFTER BRAIN LESION – REHABILITATION IMPLICATIONS: A REVIEW

Cláudia Costa[1-3,*], Ana Cristina Sequeira[3,4], Ana Cristina Inácio[3,4] and Sofia Pinto[3]
[1]Faculdade de Medicina de Lisboa
[2]Escola Superior de Saúde Atlântica
[3]Escola Superior de Saúde Egas Moniz
[4]Hospital Garcia de Orta, Lisbon, Portugal

Keywords: positive and negative neuroplasticity, brain lesion, brain recovery, physiotherapy, functional recovery

Background: Neuroplasticity has been an important thematic regarding brain lesion recovery. Neurological physiotherapists used as the basis of their clinical reasoning. However some concepts and some mechanisms within the neuroplasticity are not so well clarified in the literature and often lead to personal beliefs based on experience. The concepts of positive and negative neuroplasticity, being the positive/adaptative and the negative/maladaptative impact on functional recovery are known in semantics but really not fully understood on mechanisms and rehabilitation implications.

Objective: To conduct a literature review to investigate and describe the concepts of positive and negative neuroplasticity after brain injury, its natural recovery and the impact on physiotherapy clinical reasoning.

Methods: This review formulated research questions according to the guidelines for reviews: "What is the described as concepts of positive and negative neuroplasticity after brain injury?"; "What is the described as natural recovery after brain injury?" and "What is the implications for physiotherapy intervention?". For the search data we used the databases Pubmed, PEDro and B-on between January 2009 and January 2019 was performed with the keywords: "Positive and negative plasticity", "movement"," brain recovery"; "brain lesion"; "brain recovery"; "physiotherapy"; "functional recovery".

Results: We screened 53 studies and from these 30 studies were included in the review. Neuroplasticity it found as an important feature among the literature regarding physiotherapy intervention clinical reasoning, but only one study addressed directly to the theme of positive and negative plasticity.

Conclusion: A lot is written about neuroplasticity in general, some studies address the mechanisms after the lesion, but few approach the positive and negative concepts and its impact in rehabilitation clinical reasoning. These concepts are implied in the interventions approaches but it is important to structure and standardize, so the outcomes can be optimize. This fact will empower a qualified and specific intervention, aiming at the patient full rehabilitation potential.

In: Movement 2019: Brain, Body and Cognition
Editors: Joav Merrick and Gerry Leisman
ISBN: 978-1-53616-714-6

Chapter 162

MOVEMENT AND GESTURE ANALYSIS WITH MACHINE LEARNING IN SCHIZOPHRENIC PATIENTS

Lajos Simon, Máté Majtán**, András Horváth, Csenge Farkas and Miklós Koller
Department of Psychiatry and Psychotherapy Semmelweis University, Hungary and Faculty of Information Technology and Bionics, Pázmány Péter Catholic University, Hungary

Keywords: machine learning, schizophrenia, gesture analysis, scalability

Objective: Diagnosis of schizophrenia lacks available technologies that could enable a scalable diagnostic approach for mental illnesses with motor dysfunction. Based on previous research we hereby propose a possible diagnostic methodology paired with machine learning, which could be a visually available assisting tool for the diagnosis of schizophrenia.

Methods: Compared to currently use technological approaches, patients' motor function and gestures are first recorded and later analysed by algorithm. One patient's gestures are recorded in seven individual records. At first the patient is asked to elevate the arms to a horizontal position in the frontal axis and to turn 360 degrees around oneself creating a calibration based on the spatial data. In the next record the patient is asked to perform tasks measuring gesture. Apraxia Screen Test (AST) a test created with item reduction from Test of Upper Limb Apraxia (TULIA) is used for these tasks. The last recordings are different situation games requiring intensive emotional involvement and walking on a 8-10 meter range. The algorithm analyses the recordings, creating a database with data from patients and healthy individuals as well. Compared to the alternatives our approach requires no high-end technology, and can be performed with limited resources (space, light). The biggest achievement is the calibration algorithm that enables us to analyze spatial data without using 3D technology.

Results: The algorithm is capable of handling spatial data coming from 2D frames, and early results from 18 patients showed an easily interpretable visual display of the progress of the impaired movement and gesture performance seen in schizophrenia.

Conclusion: Based on early data we can claim, that we are iterating on a scalable diagnostic tool that is applicable in settings with limited resources. Currently we are increasing the number of patients involved in the research with national collaborations between institutions to be able to develop a better working machine learning algorithm, that could differentiate more confidently between subsets of schizophrenic patients and that could analyze progression and pharmacological efficiency of the examined patients.

In: Movement 2019: Brain, Body and Cognition
Editors: Joav Merrick and Gerry Leisman

ISBN: 978-1-53616-714-6

Chapter 163

INDIVIDUALLY GUIDED DANCE REHABILITATION MAY IMPROVE REASONING, EXECUTIVE FUNCTIONS, MOOD AND TRUNK CONTROL IN CHRONIC STATE OF TRAUMATIC BRAIN INJURY

Lilli Huttula[1,*], Jenna Leppelmeier[2], Teppo Särkämö[1], Kiki Molander[3], Kirsi Säynevirta[4], Maj-Britt Forsbom[4], Marjo Kullberg-Turtiainen[5], Petri Turtiainen[5], Jaana Sarajuuri[1], Pekka Rantanen[6], Laura Hokkanen[1], Mari Tervaniemi[7] and Sanna Koskinen[1]

[1]University of Helsinki, Faculty of Medicine,
Department of Psychology and Logopedics, Helsinki,
[2]Validia Rehabilitation Helsinki, People with Physical Disabilities, Helsinki
[3]Bodywise, Helsinki
[4]Neurotiimi TM, Helsinki,
[5]EazyeM Oy, Helsinki,
[6]Kanta-Häme Central Hospital, Hämeenlinna
[7]University of Helsinki, Faculty of Educational Sciences, Cicero Learning, Helsinki, Finland

Keywords: traumatic brain injury, dance rehabilitation, multimodal, multidisciplinary

Objective: Traumatic brain injury (TBI) is a major public health issue leading to long-term cognitive, emotional and physical impairments, calling for new and effective multimodal and multidisciplinary rehabilitation practices. Dance is a multimodal and multisensory activity that combines physical exercise and use of music, both of which have proven positive effects on healthy and neurological populations. Due to its multimodal nature, dance has the capability of activating several brain regions at the same time and, therefore, it might be ideal for enhancing similarly complex functions. The aim of our study was to develop a multidisciplinary dance rehabilitation method and to evaluate its applicability and effectiveness for TBI patients.

Methods: A total of 11 persons with severe TBI, 4 women and 7 men, participated in the study. The subjects were 19-45 years old. The average time from the injury was 7,6 years. In this cross-over study the subjects were randomized into two groups that participated in the dance intervention at different times (group A: 0-3 months; group B: 3-6 months) The intervention (3 months, two sessions per week) was carried out together by a dance instructor and a physiotherapist. Neuropsychological and motor tests were conducted at 0, 3 and 6 months. Pooled performance in general cognition, reasoning, attention and executive function, mood, quality of life, and communication were compared before and after the intervention. All significant and nearly significant results were further analyzed using pooled paired sample t-tests and standard repeated measures ANOVAs.

Results: Statistically significant results ($p < .05$) were obtained in several domains: reasoning, executive functions, mood and trunk control. Several other domains were approaching significance. Reasoning ($p = .018$), mood ($p = .005$) and trunk control ($p = .023$) were found to significantly improve during the intervention. Some executive functions had significant time*group interactions both when comparing the first two time points and all three time points.

Conclusion: This pilot study showed that individually guided dance rehabilitation may improve reasoning, executive functions, mood and trunk control in chronic state of traumatic brain injury.

In: Movement 2019: Brain, Body and Cognition
Editors: Joav Merrick and Gerry Leisman
ISBN: 978-1-53616-714-6

Chapter 164

ASSOCIATIONS BETWEEN DUAL TASKING, INNER EAR FUNCTION AND FALLS AMONG OLDER PERSONS: A LONGITUDINAL PROSPECTIVE COHORT STUDY WITH A FOLLOW-UP

Elina Valkonen*
Laakso Hospital, Helsinki, Finland

Keywords: falls, inner ear function, older persons, TUG, diffTUG

Objective: The first aim of this study was to detect whether TUG test results and differences between their values (diffTUGs) are associated with inner ear dysfunction. The second aim was to study whether TUG test results, diffTUGs and inner ear dysfunction are associated with falls. The third aim was to study whether TUG test results can predict falls.

Methods: A longitudinal prospective cohort study with a six-month follow-up was performed. TUG tests and inner ear function were tested at three balance-themed workshops for community-dwelling older persons in an exercise centre during spring 2017. Inner ear dysfunction was measured with DixHallpike test, Headshake test and Head Impulse test. Data about falls was collected with fall diaries, returned after 3 and 6 months. Logistic regressions, Mann-Whitney U-test and Fisher's exact test were used in data analysis.

Results: One hundred one women and 12 men participated. They were 64-89 years old (median 75 years, IQR 10). Of the persons tested with inner ear tests (n = 111), the majority (n = 82, 73.9%) were considered having inner ear dysfunction. Of the participants who completed the follow-up (n = 103), 25 (24,3%) fell at least once. No association was found between TUG test results or diffTUGs and inner ear dysfunction ($R2 = 0.00-0.03$). Results of TUG tests and diffTUGs were not associated with falls ($p = 0.19-0.87$) and neither was inner ear dysfunction ($p = 0.44$). Furthermore, TUG test results did not predict falls (OR = 0.88-1.16) during the follow-up.

Conclusion: The prevalence of inner ear dysfunction was higher than expected. Results of dual tasking in connection with TUG test or inner ear dysfunction did not seem to have an association with future falls. The results need to be confirmed with larger samples. Further research is needed about whether the three inner ear tests would individually be more specific predictors for falls.

In: Movement 2019: Brain, Body and Cognition
Editors: Joav Merrick and Gerry Leisman

ISBN: 978-1-53616-714-6

Chapter 165

DEVELOPMENTAL TRAJECTORIES OF NEUROANATOMICAL ALTERATIONS ASSOCIATED WITH THE 16P11.2 COPY NUMBER VARIATIONS

Alonso Cárdenas-de-la-Parra[1], Sandra Martin-Brevet[2,3,*], Clara Moreau[4], Borja Rodriguez-Herreros[2,4], Vladimir S Fonov[1], Anne M Maillard[2,5], Nicole R Zürcher[6], Nouchine Hadjikhani[6,7], Jacques S Beckmann[2], Alexandre Reymond[8], Bogdan Draganski[3,9], Sébastien Jacquemont[2,4], D Louis Collins[1] and 16p11.2 European Consortium

[1]Department of Biological and Biomedical Engineering, Montreal Neurological Institute, Montreal, Quebec, Canada

[2]Service of Medical Genetics, Centre Hospitalier Universitaire Vaudois and University of Lausanne, Lausanne, Switzerland,

[3]LREN, Département des neurosciences cliniques, Centre Hospitalier Universitaire Vaudois and University of Lausanne, Lausanne, Switzerland,

[4]CHU Sainte-Justine Research Center, Université de Montréal, Montréal, QC, Canada,

[5]Centre Cantonal Autisme, Centre Hospitalier Universitaire Vaudois and University of Lausanne, Lausanne, Switzerland,

[6]Athinoula A Martinos Center for Biomedical Imaging, Department of Radiology, Massachusetts General Hospital, Harvard Medical School, Boston, Massachusetts, Unite States of America,

[7]Gillberg Neuropsychiatry Centre, Göteborg, Sweden,

[8]Center for Integrative Genomics, University of Lausanne, Lausanne, Switzerland

[9]Department of Neurology, Max Planck Institute for Human Cognitive and Brain Sciences, Leipzig, Germany

Keywords: 16p11.2 copy number variants, neurodevelopmental disorders, genetics, brain development, normative growth trajectories

Objective: 16p11.2 copy number variants (breakpoint 4-5) confer high risk for neurodevelopmental disorders. In this study, we aim at characterizing age-related effects of 16p11.2 copy number variants by analyzing a group with a broad age range including younger individuals.

Methods: Methods used in previous studies were unable to investigate the onset of these alterations and whether they evolve with age. Here, a large normative developmental dataset was used to accurately adjust for effects of age. We normalized volumes of segmented brain regions as well as volumes of each voxel defined by tensor-based morphometry.

Results: Results showed that the total intracranial volumes, the global gray and white matter volumes are respectively higher and lower in deletion and duplication carriers compared to control subjects at 4.5 years of age. These differences remain stable through childhood, adolescence and adulthood until 23 years of age (range: 0.5 to 1.0 Z-score). Voxel-based results are consistent with previous findings in 16p11.2 copy number variant carriers, including increased volume in the calcarine cortex and insula in deletions, compared to controls, with an inverse effect in duplication carriers (1.0 Z-score). All voxel-based differences are present at 4.5 years and remain stable until the age of 23 years. Our results highlight the stability of a neuroimaging endophenotype over 2 decades during which neurodevelopmental symptoms evolve at a rapid pace.

In: Movement 2019: Brain, Body and Cognition
Editors: Joav Merrick and Gerry Leisman
ISBN: 978-1-53616-714-6

Chapter 166

MENTAL HEALTH USERS WHO CONSUME ADDICTIVE SUBSTANCES, DIAGNOSED WITH HIV IN TIJUANA, MEXICO

Ana Campuzano, Julio R Martínez-Alvarado**, Karina Pérez and Karla Vargas y Román Chávez
Universidad Autónoma de Baja California, Escuela de Ciencias de la Salud, Ensenada, Mexico

Keywords: HIV, depression, suicidal ideation

Introduction: The relationship between physical and mental health is very narrow, since there is a wide variety of diseases resulting in mental disorders, which in previous studies has been tested as one of the main risk factors in suicide and depression. Mental health problems can result in alcohol and drugs use, chronic diseases, a social and family fragile situation with isolation or social marginalization, lack of contact family, unemployment or admission to a penitentiary.

Objective: The objective of this research was to analyze the relationship between social support, depression and suicidal ideation in persons with HIV on antiretroviral drug.

Methods: The population consisted of people diagnosed with HIV attending the civil Association Hostel in the city of Tijuana, Mexico. 42 volunteers between 18-59 years of age (M = 39.85) of which 81.5% were males and drug users. A structured interview was conducted with a battery of four instruments, questionnaire on adherence to antiretroviral (SMAQ), patient health questionnaire (PHQ-9), treatment (ASIQ) suicidal ideation questionnaire and support questionnaire social functional (Duke-UNC-11).

Results: The main results indicated that there was a significant negative relationship between perceived social support, depression and suicidal ideation in people diagnosed with HIV.

Conclusion: Based on the results, we conclude that social support is an important protective factor for the mental health of people diagnosed with HIV.

In: Movement 2019: Brain, Body and Cognition
Editors: Joav Merrick and Gerry Leisman
ISBN: 978-1-53616-714-6

Chapter 167

A RETROSPECTIVE STUDY OF PRODROMAL VARIABLES ASSOCIATED WITH THE DEVELOPMENT OF AUTISM AMONG A DIVERSE GROUP OF INFANTS DURING THEIR FIRST FIFTEEN MONTHS OF LIFE

***Hanna Alonim*[1,2,*], *Giora Scheingezicht*[1], *Ido Lieberman*[3], *Danny Tayar*[1,4] *and Hillel Braude*[1]**
[1]The Mifne Center, Rosh Pinna,
[2]School of Social Science, Bar Ilan University, Ramat Gan,
[3]Department of Sociology and Anthropology, Bar Ilan University, Ramat Gan
[4]Ministry of Health, Jerusalem, Israel

Keywords: autism, children, development

Effective early intervention for the prodrome of autism within the first two years of an infant's life is contingent upon the early diagnosis of autism. This study aimed to detect the indicative developmental prodromal variables that may indicate significant risk for the later development of autism in infancy. In light of recent genetics studies highlighting the association between motor deficits in children with autism and severe spontaneous mutations, the correlation between impaired motor development and other prodromal variables identified in this study is of particular interest for us.

Methods: This study examined variables among 110 infants from all over the world diagnosed with autism at age 2-3 years. Analysis was conducted retrospectively of home videos recorded during the infants first 15 months of life. Additional developmental information was collated via parental questionnaires.

Results: Eight prodromal variables were exhibited among 89% of the infants participating in this study, including: excessive passivity (44.5%), excessive activity (28.2%), lack of eye contact (77.3%), lack of reaction to voice and presence of a parent (44.5%), refusal to eat (20.9%), aversion to touch (10%), motor development delay (33.6%) and head circumference (12.7%). Cluster analysis of combinations of variables was significant in highlighting the correlation between two variables. For example, 25.5% of the infants exhibited a combination of lack of eye contact with delayed motor development.

Discussion: The results of this study indicate that it is possible to identify early indicative prodromal symptoms associated with autism already during the first year of an infant's life. Motor development is widely recognized as an important element of early assessment, and this study confirmed that delayed motor development is an important core variable in correlation with the other prodromal behavioral variables. The variables elicited by this study provided the basis for developing the Early Symptoms of Pre-Autism Screening Scale in Infants (ESPASSI©), which is applied clinically for infants between 5-15 months. Effective application and validation of this early autism screening scale is of utility in bridging the divide between early assessment and intervention.

In: Movement 2019: Brain, Body and Cognition
Editors: Joav Merrick and Gerry Leisman

ISBN: 978-1-53616-714-6

Chapter 168

TRACKING BEHAVIORAL AND SENSORY-MOTOR DEVELOPMENTAL VARIABLES ASSOCIATED WITH THE PRODROME OF AUTISM: A LONGITUDINAL CASE STUDY OF AN INFANT'S FIRST TWO YEARS OF LIFE

Hillel Braude[1,*], Hanna Alonim[1,2], Ido Lieberman[3], Giora Scheingezicht[1], Einat Kadosh[1], Veronica Jakubson[1], Anat Ben Ari[1] and Rucha Bonnie[1]

[1]The Mifne Center, Rosh Pinna,
[2]School of Social Science, Bar Ilan University, Ramat Gan
[3]Department of Sociology and Anthropology, Bar Ilan University, Ramat Gan, Israel

Keywords: autism, infant, case study

Objective: The prodrome of autism is characterized by impairments in the emergence of behavioral precursors associated with the onset of autism. This longitudinal case study tracks the development of prodromal behaviors in an infant later diagnosed with autism through retrospective analysis of home videos filmed during his first two years of life. A key aim of the study was to track sensory-motor development in relation to other prodromal variables.

Methods: The home videos were filmed consistently by the infant's parents during his first two years of life. They were categorized in 3 month intervals from birth till 24 months and divided into twenty minute segments. Coding categories were derived from developmental behavioral variables associated with the prodrome of autism. These categories included: excessive-passivity, excessive-activity, lack of eye-contact, lack of reaction to voice and presence of a parent, refusal to eat, aversion to touch, and sensory-motor developmental delay. The variables were analyzed in terms of the intensity of the prodromal behavior correlated against a time axis, and whether the infant was in a passive or active state.

Results: Correlated together the assessment variables demonstrated three periods when the developmental deviation became increasingly obvious, i.e., 8, 14, and 20 months of age. For example, excessive passivity shifted from .45 at the age of 4-7 months to .70 at the age of 12-15 months; lack of reaction shifted from .06 at 8-11 months to .59 at 12-15 months; lack of eye contact shifted from .15 at 14 months to 2.24 at 20-24 months. The graph for sensory-motor developmental delay paralleled the other developmental variables. However, refusal to eat was distinguished from the other developmental variables in terms of more severe early manifestation and developmental trajectory (4.00 at 14 months).

Conclusion: The development of autism in this individual infant could be detected through longitudinal assessment of developmental deviation during his first 24 months. It is likely that the pattern of atypical behaviors apparent with this particular infant is indicative of the development of the prodrome of autism among infants more generally.

In: Movement 2019: Brain, Body and Cognition
Editors: Joav Merrick and Gerry Leisman
ISBN: 978-1-53616-714-6

Chapter 169

COLLECTIVE MOVEMENT ACTIVITY AS A CRITICAL PHASE IN LEARNING THE PHYSICS CONCEPT OF ANGULAR VELOCITY

Roni Zohar**, Bat-Sheva Eylon and Dor Abrahamson
Department of Science Teaching, Weizmann Institute, Rehovot, Israel and Graduate School of Education, University of California, Berkeley, California, United States of America

Keywords: embodiment, gesture, angular velocity, marking

Objective: We developed an embodied design for physics concepts. The central principle of this approach is the importance of offering students sensorimotor experiences of phenomena they are studying before analyzing them formally. We focus on the study of 10th-grade high-school physics/dance students learning the physics concept of angular velocity through participating in a collective movement activity. We seek to understand whether and, if so, learning to move together contributed to learning physics concepts.

Methods: The students were asked to solve a ‘circular group assignment’: collectively walking as one body around a bottle (a fixed point), while maintaining a straight line. Next, the instructor explained the angular velocity concept. To do so, she marked selected features of the activity by using an arm gesture (a miniaturized representation of the group’s solution of the circular group assignment). Data collected included student artifacts (e.g., drawings, creative summative projects), and transcribed video-taped sessions: the lessons, semi-structured clinical interviews, and older students’ explaining the content to younger students.

Results: Students' physical movements, developed during the collective problems, carried through to their individual interviews, teaching, and project work. These gestures appeared to anticipate verbal and symbolic utterance related to the target content. The creative summative projects manifested deep understandings of the physics concept learned while allowing students to express their personal abilities, interests, emotional affect, and philosophical involvement in the learning process.

Conclusion: Collective movement-based problem solving creates opportunities for students to ground science concepts. In particular, cognitive elements of science concepts can emerge as pragmatic solutions to the problem of coordinating the collaborative enactment of collective movement problems. This study has supported the plausibility of science curricula in which abstract concepts are designed as grounded in sensorimotor problem-solving. Moreover, our findings expand on related design research work by demonstrating the role of students' emotional, social, and philosophical involvement in their learning process.

In: Movement 2019: Brain, Body and Cognition
Editors: Joav Merrick and Gerry Leisman
ISBN: 978-1-53616-714-6

Chapter 170

FELDENKRAIS GUILD WORKSHOP: USING NON-HABITUAL CONDITIONS TO REVIVE NOVELTY AND LEARNING: EXPLORING THE PRINCIPLES AND APPLICATIONS OF THE FELDENKRAIS METHOD

Raz Ori*
Guild Certified Feldenkrais Practitioner, Ramat-Aviv Feldenkrais Center, Ramat Aviv, Israel

Keywords: Feldenkrais, learning, self-awareness, movement

Moshe Feldenkrais (1904-1984) was an Israeli physicist and judo master who developed a somatic method using sensorimotor practice meant to increase self-awareness. His approach to teaching and learning was based on a reverence for the child's way of being, valuing curiosity, exploration and spontaneity. In this workshop I aim to demonstrate and theorize about these pathways traveled during our early years through guiding seminar participants in an awareness through movement lesson, so that they may briefly return to the openness and naiveté of childhood. Many of the awareness through movement lessons Feldenkrais and his followers have developed are meant to replicate the developmental processes of infants. Replicating the process lies in following the graded complexity of the sensory-motor skills infants develop throughout the course of their first two years of life. With each emerging skill, the abilities, access, and repertoire for relating to the environment, open and expand in a variety of ways that lead to the formation of novel associations and hence novel abilities. Thus, learning is based on using prior skills, abilities and knowledge in greater complexity, provided the environmental conditions stimulate and challenge the growing infant. New opportunities in mobilization, communication and interaction have a synergistic reinforcement feedback loop that further support the child's curiosity for novel exploration. Imagine the tectonic developmental shift that takes place once a toddler has learned to crawl. Just as infants create novelty by using prior existing skills, abilities and knowledge, so do adults. Yet, as adults, our previously acquired habits may interfere and impede our attempts to learn and improve despite our best intentions. Each of us can relate to the difficulties in practicing "better sitting," triggered by the struggle to overcome our sitting habits. Thus, circumventing the dominance of the habit is a must and would entail setting up non-habitual conditions in which there is a certain amount of unfamiliarity. The unfamiliarity permits the introduction of novel ingredients into the repertoire, ingredients otherwise pushed away by the automatic pattern. These non-habitual conditions mimic the conditions in which infants develop -- conditions that provide fresh surprises, that attract the growing curiosity to explore the unknown. Using the above principles, the habitual conditions of sitting on a chair can be transformed into a "kinesthetic playground." The non-habitual sensory-motor exploration done sitting on a chair revives the kinesthetic sensations of moving and the joy of somatic self-exploration. The habitual image of sitting as a static and passive position is reshaped into a dynamic and pro-active form of self-organization.

4This workshop is supported by the Israeli Feldenkrais Guild and the International Feldenkrais Federation (IFF) in association with the IFF Research Working Group.

In: Movement 2019: Brain, Body and Cognition
Editors: Joav Merrick and Gerry Leisman
ISBN: 978-1-53616-714-6

Chapter 171

RESEARCHING THE FELDENKRAIS METHOD: A GUIDE FOR THE PERPLEXED

Hillel Braude*
The Mifne Center, Rosh Pinna, Israel

Keywords: Feldenkrais, body, movement

Objective: The Feldenkrais Method provides a systematic method for therapeutic or ameliorative engagement with a person's conscious experience (awareness) via the lived moving body. The two main methods of application include Awareness through Movement (ATM) group lessons, and Functional Integration (FI), referring to individual one-on-one hands on interaction. The Feldenkrais Method has much anecdotally reported success in alleviating pain, increasing freedom of movement, and improving general wellbeing and cognitive function. Increased public interest in the Feldenkrais Method has resulted from current advances in understanding of brain function, the development of sophisticated research methodologies into analyzing human movement, and increased public interest in somatic and mindfulness practices. Nonetheless, beyond a handful of publications, remarkably little research has been conducted into the Feldenkrais Method itself. Possible reasons for this relative deficit in researching the Feldenkrais Method include: the lack of researchers trained in the Feldenkrais Method; as well as the epistemic difficulties in devising research methodologies integrating first-person experience with third-person scientific methodologies. Clearly defined research methodologies in relation to the Feldenkrais Method have not yet been defined and rigorously applied. Addressing this deficit, in this presentation I shall review the state of the field and present guidelines for researching the Feldenkrais Method.

Methods and Results: The first part of the presentation reviews a number of different methodological approaches to researching the Feldenkrais Method based on published literature. These methodologies include: case studies, theoretical analyses such as theories of enactive perception, cybernetics, and dynamic systems approaches, neuroscience and neurophysiological research, cognitive science, health outcomes research, and movement notation, particularly the Eshkol Wachman Notation. The second part of this presentation evaluates epistemic issues in researching the Feldenkrais Method presented as a guide for junior researchers.

Conclusion: This presentation is intended to outline methodologies for researching the Feldenkrais Method that bridge the divide between various disciplines encapsulated in practicing the Feldenkrais Method, including first-person experience, cognitive neuroscience, phenomenology and holistic theories of the organism.

In: Movement 2019: Brain, Body and Cognition
Editors: Joav Merrick and Gerry Leisman
ISBN: 978-1-53616-714-6

Chapter 172

A WAY TO STUDY THE FELDENKRAIS METHOD

Eilat Almagor*[*] *and Julius Verrel
The Jerusalem Academy of Music and Dance and Center for Lifespan Psychology,
Max Planck Institute for Human Development, Berlin, Germany

Keywords: Feldenkrais, brain

Objective: Only a few studies have been done on the Feldenkrais method. One of the difficulties in designing an adequate scientific study of the method is that this method emphasizes the individual differences between people. In Functional Integration for example, the teacher is required to skillfully tune his manipulations to the student's individual behavior, and his minimal, moment to moment reactions to those manipulations. The finer the connection of the teacher to the student, in timing, direction and force, the more effective the lesson is. Individuality and subjective contact between teacher and student in effect facilitates learning, but on the other hand complicates studying it scientifically. Until one could define measurable parameters in this junction between teacher and student that consider the individual behavior of the student as well as the subjective sensing of the teacher; the "juicy" most important part of the method stays hidden. Knowing that subjectivity is indeed in the heart of the method, we challenged ourselves to establish a scientific study of the effect of a subjective image that the teacher creates in his mind about his student, on the students' brain activity. This presentation analyzes the methodology of our study in terms of the fundamental question of determining a scientific way to study the Feldenkrais method.

Methods: Twenty-one participants participated in the study. Participants were in a supine position in an MRI scanner with extended legs, while an experienced Feldenkrais practitioner used a planar board – an artificial floor – to touch and apply minimal force to different parts of the sole and toes of their left foot under two experimental conditions. In the local condition the teacher explored built a dynamic image of the foot, toes and ankle and in the global condition the teacher imagined the movement from the foot up the leg and torso towards head.

Results: The grouped results over 21 subjects showed a difference in the resting students' brain activity whether it was after being manipulated in the local case versus being manipulated in the global case.

Conclusion: The study approach can be useful for future research of the Feldenkrais method.

In: Movement 2019: Brain, Body and Cognition
Editors: Joav Merrick and Gerry Leisman
ISBN: 978-1-53616-714-6

Chapter 173

A NEUROPLASTICITY SCALE FOR ASSESSING THE FELDENKRAIS METHOD FOR MS PATIENTS BASED ON INDIVIDUAL REPORTS

Triantafillia Reziti*
Feldenkrais Praxis and Somatics Center, Perama, Greece

Keywords: Feldenkrais, multiple sclerosis

Objective: My study is focused on understanding and solving complex clinical issues that disturb patients with multiple sclerosis (MS). My goal was to utilize an inductive methodology which could transform the subjective experience into a general rule, rather than trying to fit subjects into existing research categories. This necessitated the development of an effective and easy-to-use evaluation tool, based on the Feldenkrais Method that addresses multiple qualities, which have not been captured simultaneously for MS patients. The proposed tool – called Neuroplasticity Scale (NS) – enables both therapists and clients to assess the changes in sensory-motor and cognitive processes during action.

Methodology: A MS patient with severe instability and spasticity was selected as a case study in order to analyze the developed methodology including: a) intervention and b) home training. The intervention consisted of a six week training designed according to the symptoms of the subject and had three parts: i) discussion – evaluation by NS, ii) Functional Integration, and iii) training, which was recorded and used for motor learning comprising of auditory cues for perception, movement patterns, spatial orientation, movement timing, changes of attention and analysis of everyday movement sequences.

The NS evaluation was performed before and after the intervention, thus aiding the elucidation of the missing cognitive links and guiding subsequent practice.

Results: The procedure of the home training enabled the subject's engagement and led to improved balance and reduced spasticity. The NS proved to be a useful tool to evaluate the variables that define the clinical picture of the subject's present abilities and deficits. This methodology combined with the practitioner's observational and interactive skills may lead to a fine tuning of the home training in order to match the individual's functional needs.

Conclusion: This approach was developed to confront a disease with a multitude of symptoms that result in a clinical picture that often does not correspond to the results acquired by advanced imaging techniques used to evaluate the progress of the disease. The NS offers a useful clinical, as well as a possible research tool that can be further tested for reliability and validity with MS patients.

In: Movement 2019: Brain, Body and Cognition
Editors: Joav Merrick and Gerry Leisman
ISBN: 978-1-53616-714-6

Chapter 174

EXECUTIVE FUNCTIONS AND AUDIO-MOTOR SYNCHRONIZATION IN ATHLETES

Anastasia V Kovaleva*
Anokhin Research Institute of Normal Physiology,
Moscow and Sports Center of Innovative Technologies and Teams Exercise Training,
Moscow Sports Committee, Moscow, Russia

Keywords: athletes, audio-motor synchronization, executive functions, rhythm

Objective: Recent researches showed that performance in the audio-motor synchronization task (both accuracy and variability) is associated with fine executive functions characteristics: spatial working memory, high processing speed and visual spatial span, attention and inhibition. The aim of the current research is to reveal some associations between executive functions and rhythm perception and reproduction in healthy adults.

Methods: Forty five athletes (30 males and 15 females, mean age 19.4 ± 4.1) of different kinds of sports participated in the experiment. For executive functions assessment, two tests from the Viena Test Systems (Austria) was used: Corsi test for spatial working memory measurement and Stroop test for inhibition measurement. For attention shifting concentration grids were used ("Praktika-MSU"). During audio-motor synchronization tasks athletes had to tap out a regular beat with their dominant hand on a special pad. They performed four simple synchronization tasks: to synchronize hand movements (taps) with the external sounds with 40, 60, 90 and 120 bpm frequencies. Then participants performed "rhythm memory test" and "interfering rhythm test" (designed for distractibility assessment). The rhythm memory task consisted of two parts: the synchronization phase (tap according to the metronome sound) and a continuation phase (without auditory cue). The interfering rhythm task also consisted of two parts: 1st - to tap accordingly to the 60 bpm metronome beat (30 seconds), 2nd - additional rhythm with different frequency and sound was switched on, and an athlete should continue to tap according to the basic 60 bpm rhythm (next 30 seconds) and not to pay attention to the interfering rhythm (distractor).

Results: Associations were revealed between audio-motor synchronization and Stroop test. It was also shown that more accurate and less variable in rhythm tests athletes demonstrated better results in shifting test. Despite the working memory test assessed visual spatial abilities, higher working memory capacity was related to less variable audio-motor synchronization.

Conclusion: Correlation analysis revealed close relations between rhythm synchronization task results and Stroop test, attention shifting and working memory performance. The better audio-motor synchronization, the better executive functions are.

In: Movement 2019: Brain, Body and Cognition
Editors: Joav Merrick and Gerry Leisman
ISBN: 978-1-53616-714-6

Chapter 175

CORRELATION OF A QUANTITATIVE MRI SCORING SYSTEM WITH MOTOR FUNCTION IN CHILDREN WITH CEREBRAL PALSY

Mitchell Schertz[1-3,*], Shelly I Shiran[1,4,5], Miriam Tsveyer[3], Liora Sagie[1,3], Sharona Katzenellenbogen[3], Daphna Domanievitz[3] and Aviva Fattal-Valevski[3,5]

[1]Child Development and Pediatric Neurology Service,
Meuhedet- Northern Region, Haifa, Israel
[2]Faculty of Medicine, Technion, Haifa, Israel
[3]Pediatric Neurology Institute, Tel Aviv Sourasky Medical Center, Tel Aviv, Israel
[4]Pediatric Radiology Institute, Tel Aviv Sourasky Medical Center, Tel Aviv, Israel
[5]Sackler Faculty of Medicine, Tel Aviv University, Tel Aviv, Israel

Keywords: cerebral palsy, motor function, GMFM, GMFCS, MACS

Objective: Children with cerebral palsy (CP) present with variable degrees of motor impairment that can be assessed using standardized instruments including the Gross Motor Function Measure-66 (GMFM-66), Gross Motor Function Classification System (GMFCS) and the Manual Ability Classification system (MACS). Magnetic resonance imaging (MRI) is used to identify etiology and extent of insult to the central nervous system in infants and children presenting with symptoms of CP. However, the findings are often descriptive. Recently, a novel coding system based on quantitative assessment of abnormalities seen on the MR images, was reported to correlate with measures of motor function in a cohort of children with hemiplegic CP. The objective of this study was to assess the correlation of a formalized coding system with clinical parameters of motor dysfunction in children with different types of CP.

Methods: This cross-sectional study included children diagnosed with CP who had an MRI scan performed after age 18 months. For this study, children underwent measures of motor function including GMFM-66, GMFCS and MACS. Patients' MRI studies were coded using a previously validated measure to produce a radiological score (RS). Results were analyzed to identify correlations between clinical and imaging parameters.

Results: The cohort consisted of 35 children (19 male, 16 female) aged 9.1 ± 3.7 years. Types of CP included quadriplegia 14 (41%), hemiplegia 12 (35%), diplegia 7 (21%) and triplegia 1 (3%). Etiology of CP included periventricular leukomalacia (PVL), intraventricular haemorrhage (IVH), stroke, hypoxic ischemic encephalopathy, with the majority of children having two etiologies, 66% having PVL and 43% having IVH. Age of walking for the cohort was 1.9 ± 1.3 years. GMFM-66 score 67.5 ± 33.1. GMFCS score 2.6 ± 1.5. MACS score 2.5 ± 1.3. RS was found to be correlated with GMFM-66 (r-.547, $p < 0.01$), GMFCS (r-.461, $p < 0.01$) and MACS (r-.609, $p < 0.01$).

Conclusion MRI-based radiologic scoring system for extent of brain injury in children correlated with measures of motor function in children with CP.

In: Movement 2019: Brain, Body and Cognition
Editors: Joav Merrick and Gerry Leisman
ISBN: 978-1-53616-714-6

Chapter 176

THE SELF-KNOWING BODY AS AN INSPIRATION FOR DANCE

Michal Shahak*
International Trauma-Healing Institute, Raanana, Israel

Keywords: workshop, movement, quality, cognition

In this experientially based, somatic-movement workshop, participants are engaged in an active practice that enhances movement articulation and enriches the participant's quality of movement. The workshop demonstrates an experiential learning process whereby human anatomy is learned from within in a mode of embodied cognition, which in turn becomes an inspiration for expressive movement and dance.

The practice is comprised of a sequence of three phases: cognitive learning of structural anatomical information; 'implementing' that information as a somatic-image, i.e., as an action of visualization in the body; and finally, initiating movement from the experience of somatization. This practice facilitates an inner knowing of the body as an embodied practice that engenders new qualities of movement, deeper and reaching beyond the dancer's formal training. It enables the body's self-knowledge to become a mine for dance creation. This suggests a new discipline, based on movement that unfolds from within.

This new discipline becomes a movement language. It trains dancers to become attentive to the raw micro-movements that underlie grosser movements. Analytically, this discipline follows the scientific principle of decomposing complex wholes into elementary particles as a first step towards recomposition and further development. The unraveling of movement into its inwardly sensed constitutive parts becomes the basis for kinesthetic expression, in which the body learns to apperceive itself through the experiential mastering of its structures.

In: Movement 2019: Brain, Body and Cognition
Editors: Joav Merrick and Gerry Leisman
ISBN: 978-1-53616-714-6

Chapter 177

PHYSICAL SOMATIC IMAGES VS EMOTIONAL IMAGES AS A PLATFORM FOR A SOMATIC EXPERIENCE AND IMPROVISED DANCE

Michal Shahak*

International Trauma-Healing Institute, Raanana, Israel

Keywords: workshop, movement, quality, cognition Michal Shahak presenter

Movement and dance are initiated and created from many different sources and resources. Inspiration for movement and dance-creation is vast – starting from memories, emotional content and context, daily experiences, social relationships, political subjects and aspects, stories and narratives, bodily sensations, bodily memories or content and bodily context.

As human beings we are psycho-physical entities. The relationships between body and mind are interrelated and mutually co-operate on and within each other's sphere. And yet it's a fruitful exploration - a kind of "dance", actually - to question and investigate the way emotional images stimulate the dancer/mover vs the way physical and somatic images do. Issues addressed are, for example: the differences in the qualities of dance and the way it is experienced, engendered by the two sources, as well as the different action that the emotional image vs the physical image have on the body and its movement.

The lecture invites the audience to explore questions on a bodily/sensational level while listening to it. It discusses the body as the field of manifestation for any emotional image, thought or feeling. Initial physical and somatic images as a source for movement and bodily action are considered and suggest exploration as well. The language of internal sensations is explored as a manifestation of feelings, emotions and thoughts. This language of sensations is also examined as the source for movement in and of itself.

In: Movement 2019: Brain, Body and Cognition
Editors: Joav Merrick and Gerry Leisman
ISBN: 978-1-53616-714-6

Chapter 178

INVESTIGATING LANGUAGE DENSITY EFFECT ON EYE MOVEMENT BEHAVIOR (RUSSIAN VS ENGLISH)

Valeriia Demareva *
Department of Psychophysiology, Lobachevsky State University,
Nizhny Novgorod, Russia

Keywords: eye tracking, reading, language density

Objective: The objective of the study was to investigate patterns of global eye movement measures in native Russian reading in order to create a better understanding of language density effect on eye movement behavior. We assume that the information density of words in Russian is less than in English, due to its longer words. This is supposed to result in greater number of fixations and larger saccades, as well as in less duration of fixations when reading equivalent texts in Russian versus English.

Methods: A group of 27 Russian students with normal or corrected to normal vision took part in the experiment. The participants demonstrated an equally high level of Russian according to C-test. Eye movements were recorded binocularly with an SMI Hi-Speed tracker at 500 Hz. After passing a 9-point calibration, participants read eight texts and answered comprehension questions. The texts were a Russian-translated version of the stimuli in (Liversedge et al., 2016). The texts were divided into 2-6 slides, each of which contained 1-8 sentences, and one character subtended 0.46 visual angle.

Results: The sentence was taken as a unit of analysis, and four eye movement measures were computed: total sentence reading times, average number of fixations, average forward saccade size, and average saccade duration. Data more than 2.5 standard deviation from the mean for each participant and fixations shorter than 60 ms or longer than 800 ms were removed from the analysis. As compared to the results obtained for English (Liversedge et al., 2016) in Russian we observed longer sentence reading times (Mrus = 4302 ± 1865 ms; Meng = 3093 ± 777 ms), less number of fixations (Mrus = 8.6 ± 2.38; Meng = 14.81 ± 2.93), shorter saccade size rightwards in characters (Mrus = 7.78 ± 1.79; Meng = 8.53 ± 1.55) with shorter fixation duration (Mrus =195 ± 23 ms; Meng = 207 ± 32 ms). There is a significant effect of text type, e.g., on total sentence reading times ($F (6, 3701) = 27.5$, $p < 0.001$).

Conclusion: The results confirm the assumption about fixation duration in Russian native reading vs. English, but contradict it about shorter saccades and more fixation count suggesting that eye movements in silent reading were effected by specific linguistic features of Russian (besides density).

Supported by RFBR (grant N 18-013-01169).

In: Movement 2019: Brain, Body and Cognition
Editors: Joav Merrick and Gerry Leisman ISBN: 978-1-53616-714-6

Chapter 179

THE LINK BETWEEN GAZE FOLLOWING BEHAVIORS AND EARLY LANGUAGE

Lea Lasman*

Department of Communication and Disorders, Adelphi University,
Garden City, New York, United States of America

Keywords: gaze following, language, social cognition, theory of mind

Objective: Learning about objects in the environment would be impossible without gaze following. The eye gaze and subsequent orientation of the head and body indicates an observation made by one individual and directs the attention of another to a salient object. Certain salient features of gaze following are predictors of the mental states of others, and in humans are reflective of the theory of mind (ToM). In children, the ToM allows the representations of mental states of others and can be an early predictor of symbolic thinking and language function later in life. Infants track the movements of an adult's eye and head instinctively, and develop the ability to look in the direction indicated.

Methods: In developing, pre-linguistic children, determining the precise direction of the attention using eye movements without another individual provides salient information about the child's understanding of the location of objects and may also convey more complex social information.

Results: [results pending] The observation of animal versus an object that does not require another individual's direct attention is measured. Systematically showing any rudimentary movements are innate.

Conclusion: Gaze following provides important cues about the location of objects, in many animals. This functions as a complex form of social cognition.

In: Movement 2019: Brain, Body and Cognition
Editors: Joav Merrick and Gerry Leisman

ISBN: 978-1-53616-714-6

Chapter 180

LATE-LIFE MOTOR IMPAIRMENTS: AN UNDERSTUDIED ALZHEIMER DISEASE PHENOTYPE

Aron S Buchman*
Rush Alzheimer's Disease Center, Rush University Medical Center,
Chicago, Illinois, United States of America

Keywords: Alzheimer's disease, ADRD, motor, impairment

Objective: Alzheimer's disease and other related disorders (ADRD) may manifest non-cognitive phenotypes. To what extent is late-life motor impairment related to ADRD phenotypes.

Methods: We used data from older adults from three community-based longitudinal cohort studies [Religious Orders Study, Rush Memory and Aging Project, and Minority Aging Research Study]. Participants were enrolled without known dementia and agreed to annual clinical evaluations and autopsy. All three studies share a large common core of testing batteries including uniform annual structured testing of cognition and several motor phenotypes and collection of postmortem ADRD indices, facilitating combined analysis.

Results: We examined the onset of AD dementia and mobility disability (gait speed < 0.55m/s) in 1262 adults unimpaired at baseline. Mobility disability preceded AD dementia in 74%, AD dementia occurred first in 12% and both disabilities occurred together in 14%. Impaired mobility with normal cognition suggests that there may be distinct CNS drivers of motor impairment. Next we used a linear mixed effect model in 1430 decedents to determine if ADRD indices are related to the progression of

Parkinsonism, a common aging motor phenotype. The following brain pathologies including AD, TDP-43, Parkinson's disease pathology, macro infarcts, atherosclerosis, arteriolosclerosis, and CAA were independently related to progressive parkinsonism. While these brain indices may account for up to 50% of the variance of cognitive decline, in this model they accounted for < 20% of the variance of progressive Parkinsonism. In a subset of decedents (N = 250) ADRD indices such as tangles and Lewy bodies extend beyond the brain to all levels of spinal cord. While spinal indices explain additional variance of motor function when controlling for brain indices, the majority of motor function remained unexplained.

Conclusion: To determine the degree to which ADRD indices drive motor impairment, indices must be obtained from the full extent of the motor control systems which extend outside the brain to reach peripheral muscle. As ADRD indices may only partially explain motor impairment, molecular mechanisms without known pathologic footprints remain to be identified. Finally, the use of wearable sensors may allow investigators to capture more facets of motor function to facilitate studies of this complex phenotype.

Funding: This work was supported by National Institute of Health [P30AG10161, R01AG15819, R01AG17917, RF1AG22018 R01AG44379, R01NS78009, R01AG47 976, R01AG56352, R56AG59732, R01AG59732]

In: Movement 2019: Brain, Body and Cognition
Editors: Joav Merrick and Gerry Leisman
ISBN: 978-1-53616-714-6

Chapter 181

DUAL TASK ASSESSMENT DURING TREADMILL WALKING FOR CHILDREN AND ADOLESCENTS WITH SPORT CONCUSSION

Nour Saade[1,2], Karine Boivin[1,2], Zachary Beaudet[1,2] and Philippe Fait[1-4,*]

[1]Department of Human Kinetics; Université du Québec à Trois-Rivières (UQTR), Trois-Rivières, Quebec, Canada
[2]Research Group on Neuromusculoskeletal Dysfunctions (GRAN), UQTR, Trois-Rivières, Quebec, Canada
[3]Research Center in Neuropsychology and Cognition (CERNEC), Montréal, Quebec, Canada
[4]Center for Interdisciplinary Research in Rehabilitation and Social Integration (CIRRIS), CIUSSS-CN, Quebec city, Quebec, Canada

Keywords: concussion, treadmill, walking, children, sport

Concussions may be arduous to assess due to the multiple systems affected and the assortment of symptoms subtle to detect. Literature suggests that some gait parameters are sensitive to brain dysfunctions. This study interest is to determine which parameters are pertinent in a cognitive-motor task and its anticipatory effects. The objective was to identify sensitive gait parameters (GP) relevant to detect functional and cognitive dysfunctions in youth who have sustained a sport concussion.

Methods: Youth aged between 10-17 years old ($n = 20$, mean age $= 13.8 \pm 1.7$ years) took part to this study. Treadmill walking (TW) was assessed in single or dual task conditions. The second task could be cognitive (modified Stroop task, STROOP) or motor (stepping over an automated dynamic obstacle at 0.115 m above the treadmill surface). Foot and obstacle kinematic data were recorded with an optical tracking system. The STROOP was displayed on a digital tablet placed in front of participant. Five specific conditions were tested in this trial order: (1) TW (single task condition), (2) TW while waiting for STROOP task (anticipation of a cognitive dual tasking), (3) STROOP task concomitant to TW (cognitive dual tasking), (4) TW while waiting for the obstacle launching (anticipation of a motor dual tasking) and (5) Stepping over the obstacle during TW (motor dual tasking). ANOVAs were done to determine condition and group effects on specific GP: stride length and duration as well as different variables issued of vertical and anteroposterior foot obstacle clearance. Two separate analyses were done on mean values of these GP and on their inter-gait cycle variability.

Results: We observed condition effects, for the 2nd and 4th conditions (involving the anticipation of a cognitive or a motor dual tasking) on the stride inter-gait cycle variability, in particular. This effect was more obvious in the concussed than healthy youth. Concussed youth should be more affected by the dynamic obstacle (5th condition) by shortening their vertical foot obstacle clearance.

Conclusion: This work could be relevant to detect sensitive gait parameters to the effects of concussions in a cognitive-motor dual tasking for a better assessment and follow up of children and adolescents. This research project have been supported by a research grant from FRQ-S (governmental agency).

In: Movement 2019: Brain, Body and Cognition
Editors: Joav Merrick and Gerry Leisman
ISBN: 978-1-53616-714-6

Chapter 182

SPACING THE STAGE: THE INNER STAGE OF THE DANCER

Vered Aviv*
The Jerusalem Academy of Music and Dance, Jerusalem, Israel

Keywords: dance, imagery, space perception, mental representation

Objective: Dancers learn choreographed dances, practice and rehearse them for many dozens of hours in a specific studio space. Yet, they perform these same dances, with high precision level, in different stages and after a short-term rehearsals (minutes to a couple of hours) in the new stage/space. How do dancers adjust to accurately perform the same dance in frequently changing stages?

In order to correctly behave in a given physical space, dancers need to develop a cognitive map of this space. This map includes the position ("where am I" e.g., using hippocampal "place cells"), the dimensions (a "ruler", using "grid cells") and the boundaries of that space ("boundary cells"), as well the direction of their head, ("head direction cells"). Such an internal cognitive 3D map of a newly learned space is constructed following an intense physical exploration of that space. Dancers often find themselves performing in very different spaces (stages) to which they must adjust rather fast, so that they could perform precisely a particular sequence of movements that they have practiced in their familiar space/stage.

Methods: The present study is based on a set of questionnaires which was provided to professional dancers and semi-professional dancers, as well as interviews with choreographers aiming to explore the process that they undergo during adjustments to a new performing space.

Results and Conclusion: The present work introduces the idea that, via imagery, professional dancers build in their mind an “inner dancer’s stage” while learning and rehearsing a dance. This inner reference space has particular dimensions and structure. Later, when performing at different locations, this inner dancer’s stage is mentally “carried” over by the dancer to the new space. Depending on the rehearsal time at the new stage, the dancer partially (or fully) adapts his/her inner stage to the actual space and partially (or not) dance in relation to his/her previously built inner stage.

In: Movement 2019: Brain, Body and Cognition
Editors: Joav Merrick and Gerry Leisman
ISBN: 978-1-53616-714-6

Chapter 183

PROXIMAL VERSUS DISTAL UPPER EXTREMITY MOTOR IMPAIRMENTS POST-STROKE

Sharon Israely**, *Gerry Leisman and Eli Carmeli
Department of Physical Therapy, Faculty of Social Welfare and Health Sciences,
University of Haifa, Haifa, Israel

Keywords: stroke, Fugl-Meyer, elecromyography, handwriting

Objective: To evaluate the extent of damage to handwriting skills compared to arm reach and grasp task among post-stroke patients.

Methods: Eighteen patients and 19 healthy subjects were recruited to this case-control study. Patients were evaluated 15.2 days (± 6.5) after the stroke using a Computerized Penmanship Evaluation Tool, surface Electromyography and Fugl-Meyer assessment. This study compared motor deficits in hand reaching and grasping and in handwriting between stroke patients and healthy subjects. Damage to handwriting performance relative to hand reaching and grasping skills was also evaluated.

Results: Significant differences were found between groups in handwriting performance ($p < 0.05$). The performance of the trapezius, biceps, and triceps muscles can predict 63.5% of the variance in the ability to write a short sentence ($p < 0.023$). Pen pressure can predict 74.9% of the hand motor performance from Fugl- Meyer assessment ($p < 0.05$). Handwriting was more damaged than was the pattern of activation of the proximal muscles of the shoulder and arm ($p < 0.05$). FM scores were highly, negatively correlated with the in-air writing time across tasks ($r = -0.819$, $p < 0.004$).

Conclusion: This study confirms the clinical observation that dexterity skills are more damaged than are arm forward reach after a stroke. This possibly might be related to damaged corticospinal projections to distal musculature and larger cognitive demands during handwriting. However, these differences in motor performance were not significant in mildly disabled patients, demonstrating the feasibility of handwriting rehabilitation in these patients. Therefore, we modestly recommend focusing on handwriting rehabilitation of the hemiparetic upper extremity in mildly impaired patients after a stroke.

In: Movement 2019: Brain, Body and Cognition
Editors: Joav Merrick and Gerry Leisman
ISBN: 978-1-53616-714-6

Chapter 184

CHANGES IN RESPIRATORY EXCURSION AND STANDARD QUALITY OF LIFE MEASURES FOLLOWING PHYSICAL TRAINING FOCUSED ON PROPRIOCEPTION AND FLUID FASCIAL MOVEMENT

Satya Sardonicus* ***and Danielle Ayres***
Independent clinical practice, Portland, Oregon, United States of America

Keywords: fascia, respiration, stress, sympatheticotonia, movement

Objective: Many individuals with chronic stress experience difficulty with deep breathing. Fascial adhesions around the ribcage reduce the depth of respiratory excursion. Deep breathing can help mediate the effects of stress and is required to optimize a multitude of physiological processes. Any mechanical restrictions reduce normal proprioceptive signaling to the brain, and instead send nociceptive signals which upregulate a sympathetic response. When restrictions are present for an extended period of time, this can contribute to sympatheticotonia, or a chronically elevated sympathetic response. Chronic stress plays a significant role in many health challenges, and the brain's ability to handle stressors of all kinds may be compromised in the presence of persistent nociceptive signaling. Sympatheticotonia is associated with lower overall scores on standardized quality of life measures, including pain levels, anxiety, depression, fatigue, sleep quality, social participation, and other activities of daily living. The goal of this study was to measure cognitive, quality of life, and physiological changes following a movement training and self-therapy protocol designed to encourage fluid motion throughout primary fascial structures, including but not limited to direct work on the ribcage.

Methods: Individuals underwent a series of movement and self-therapy training modules where they performed self-fascial release manual therapy and fluid movement exercises promoting proprioception. Respiratory excursion was measured by lower and upper ribcage circumference at full expiration and full inspiration. These measurements were taken before and after training. Participants also filled out the SF-12 questionnaire before and after training.

Results: On average, participants demonstrated an 80% increase in depth of respiratory excursion, with some participants gaining as much as a 2.25 inch difference in ribcage circumference during full expiration versus full inspiration. Quality of life measures improved by an average of 80%.

Conclusion: This study suggests that individuals displaying restriction in depth of respiratory excursion and challenges in quality of life measures may benefit from a movement-based exercise and self-therapy program focused on reducing fascial adhesions and activating additional proprioceptive signaling through fluid movement. Future studies should explore additional markers of sympatheticotonia and the impact of similar movement-based therapies on autonomic balance.

In: Movement 2019: Brain, Body and Cognition
Editors: Joav Merrick and Gerry Leisman
ISBN: 978-1-53616-714-6

Chapter 185

DANCER'S SOMATIC OF MUSICALITY

Niv Marinberg [*] ***and Vered Aviv***
The Jerusalem Academy of Music and Dance, Jerusalem, Israel

Keywords: dance, performance, rhythm, melody, choreography, movement, music synchronization

Objective: Dancers often perform while synchronizing their movements to music as required by the choreographer. In this study, we introduce the concept of categorizing choreography (or segments of it) according to its relationship to either the rhythm or the melody of the accompanied music, or to both. We ask whether professional dancers tend to synchronize their movements better to either the rhythm or to the melody of the music. We wonder whether such putative differences impact dancers' capability to learn or to perform certain dance pieces.

Method: In a pilot study, we composed choreographic sequences that were either melodic-based or rhythmic-based and taught them to professional dancers. After a week time, the dancers performed the previously learned sequenced, one after the other, in front of a camera. The performed-filmed choreographic pieces were evaluated according to the timing of the movements with respect to the music.

Results and Conclusion: Our research demonstrate that some dancers tend to synchronize their movements better to the rhythm and others to the melody. We refer to this tendency as the "dancers' somatic of musicality". The findings highlight important differences in the somatic of musicality among dancers and call for attention of both choreographs and dancers to these differences since they bear on the way dancers learn, memorize and perform.

In: Movement 2019: Brain, Body and Cognition
Editors: Joav Merrick and Gerry Leisman
ISBN: 978-1-53616-714-6

Chapter 186

AN APPROACH TAILORED FOR SPECIAL NEEDS STUDENTS: SHALL WE DANCE?

Sean Ziv*
Dance Tel Aviv Studio, Tel Aviv, Israel

Keywords: dance, Parkinson, autism, movement, motor skills

Background: It is widely held that there is compensation mechanism of the brain when the individual is under pressure and forced to deal with issues that were supported by the injured or malfunctioning area. The levels of serotonin and vasopressin are also believed to increase through laugher and human touch

Objective: To improve motor and coordinative skills in people with Parkinson's disease (PD), Multiple Sclerosis (MS) and those on the autistic spectrum, as well as enhancing their social skills, and address issues of depression and poor self-image.

The method: Teaching couple dancing in the areas of Latin (salsa, bachata, rumba etc.), Ballroom (waltz, foxtrot etc.) and Argentine Tango through a structured methodology and tools.

Anticipated results: 1) Improvement in motor and coordinative skills, heightened awareness of their surroundings and application of the dance teaching methods into everyday life especially for individuals with PD and 2) improvement of social skills, especially for young individuals with autistic spectrum disorders.

Discussion: Highly rewarding results in areas of motor, coordinative, cognitive and communication skills have been achieved by following structured training in social dancing. Dance instructors of people with special needs, require clear-cut lesson plans tailored to current anticipated levels of the student's abilities.

In: Movement 2019: Brain, Body and Cognition
Editors: Joav Merrick and Gerry Leisman
ISBN: 978-1-53616-714-6

Chapter 187

THE PROCESS OF SPACE PERCEPTION: HOW DO WE PERCEIVE ARCHITECTURE?

Avishag Shemesh**, Gerry Leisman, Ronen Talmon, Moshe Bar and Yasha (Jacob) Grobman

Faculty of Architecture and Town Planning, Technion, Haifa, Israel
Faculty of Health Science, University of Haifa, Haifa, Israel
Faculty of Electrical Engineering, Technion, Haifa, Israel
Brain Research Center, Bar Ilan University, Ramat Gan,
Faculty of Architecture and Town Planning, Technion, Haifa, Israel

Keywords: Space, perception, architecture

The effect of the built environment over the user is a growing area of study, as it is important to architects and initiators to plan environments suited to the potential users. Yet, how we perceive architecture remains vague. Using new eye-tracking devices and virtual reality capabilities, we can investigate the process of architecture perception. In ongoing research, we follow the process of space perception among users, both designers and non-designers, in an immersed virtual environment.

In our study, we were able to isolate the aspect of geometry. Our developed methodology includes several physiological sensors while eye-tracking sensors indicate specific focus participants possess in a given moment. Participants physiological and mental response was traced, analyzed and compared to their pupil's movement and dilation, to determine their preference, excitement and areas of interest.

According to preliminary findings, we seem to recognize patterns in which participants perceive spaces. We believe this research may contribute to our understanding of the effect architecture has over the user. This may encourage designers to become more aware of their design potential and create architecture, which fit best to the potential user.

In: Movement 2019: Brain, Body and Cognition
Editors: Joav Merrick and Gerry Leisman

ISBN: 978-1-53616-714-6

Chapter 188

ACQUIRED ACRODERMATITIS ENTEROPATHICA AFTER ENTERAL NUTRITION IN AN ELDERLY FEMALE

Julie Bittar**, Allison Foster, Noor Bittar, Theresa Rohr-Kirchgraber and Kristine Hess
Medical Student Council, Indiana University School of Medicine,
Indianapolis, Indiana, United States of America

Keywords: acrodermatitis, enteral nutrition, case report

A 72-year-old female presented to dermatology clinic with a two month history of progressive rash on her scalp, lips, oral mucosa, trunk, and extremities. One day prior, she had a feeding tube removed that was in place for three months for dysphagia. She reported severe weight loss from poor eating, but was otherwise asymptomatic. She had no recent medication changes. On exam, her scalp, arms, chest, back, legs, feet and periorificial area had large ill-defined and well demarcated red, scaly thin plaques. She also had erosions on vaginal lips, and erythema in the perianal area and oral mucosa. Erythematous macules with peripheral scale were noted on the feet. Laboratory evaluation showed zinc-deficiency with a serum zinc level of 551 mcg/L (Reference Range 700-1200 mcg/L), while ferritin, 25-OH-vitamin D, and Vitamin B12 were within normal limits. A diagnosis of acquired Acrodermatitis Enteropathica (AE) was made and treatment was initiated with zinc gluconate. Zinc levels returned to normal limits and skin lesions promptly resolved.

Conclusion: We present an unusual case of acquired Acrodermatitis Enteropathica in an elderly female on tube feeds. Although acquired acrodermatitis enteropathica has been reported in patients on total parenteral nutrition, this is one of the first cases reporting this disease after tube feeds.

Clinical significance: Zinc supplementation is especially important in the elderly, as this population is at an increased risk of zinc deficiency due to decreased intake and impaired absorption. Elderly women are particularly at risk for zinc deficiency, and accordingly, for acrodermatitis enteropathica. Acrodermatitis Enteropathica is a disorder of zinc metabolism, resulting from either impaired absorption or deficient intake, and manifests as erythematous, desquamative dermatitis, cheilitis and diarrhea. Dysphagia, as seen in our patient, is a significant problem in elderly women that is contributory to vitamin deficiency and indication for alternative feeding modalities (tube feeds, total parenteral nutrition). Furthermore, studies have shown that defective absorption of zinc in the elderly that may indicate a need for higher nutrient requirements than younger patients. The leading theory is that defective metallothioneins and transport proteins in the elderly can result in decreased intracellular zinc. The current recommendation for zinc is 8 mg for all women over 19 years. However, this current requirement may not be enough given that studies show a significant portion of elderly women are deficient in zinc.

In: Movement 2019: Brain, Body and Cognition
Editors: Joav Merrick and Gerry Leisman
ISBN: 978-1-53616-714-6

Chapter 189

MOTOR AND COGNITIVE ASPECTS OF SEXUALITY

Gila Bronner*
Sex Therapy Services, Sheba Medical Center, Tel Hashomer, Israel

Keywords: sexuality, intimacy, motor, cognitive, parkinsonism

Objective: Sexuality is a complex multifaceted process that requires coordinated functioning of the person's cognitive, emotional, autonomic, sensory, and motor systems. In parkinsonism many aspects of intimacy and sexuality can be disrupted and impact negatively the quality of life of patients and their spouses. The objectives of this presentation are 1) to increase the awareness of health care professionals to the interplay between movement, cognition and patient's sexuality; 2) to understand the role of health professionals in discussing these underdiagnosed issues; 3) to offer practical interventions in coping with patients' intimate and sexual problems.

Methods: Analysis of the basic motor and cognitive aspects needed for intimacy and sexuality in Parkinsonism, was based on clinical experience in sexual counseling of over 30 years in movement disorder clinics.

Results: Sexuality was divided into two types of activities: (1) non-erotic and intimate activities, (2) erotic sexual activities. The role of motor and non-motor symptoms in disturbing these two areas of sexuality was evaluated. It was found that motor as well as cognitive problems effected the ability to express love and intimacy (touching, hugging, kissing, cuddling, finding comfortable and stable positions, concentration problems, communication and verbal problems to express emotions). Problems in fine motor movements, tremor, rigidity, inability to find proper positions for intercourse, disturbances in attention and imagination interrupt normal sexual response and resulted in erotic sexual dysfunctions. These changes increase patients' anxiety and couple discord.

Conclusion: Previous studies show that patients highly appreciate sexual discussions initiated by their health care professionals. Providing simple explanations and limited information were found to decrease anxiety of patients and their spouses.

In: Movement 2019: Brain, Body and Cognition
Editors: Joav Merrick and Gerry Leisman
ISBN: 978-1-53616-714-6

Chapter 190

MAPPED TAXONOMIES FOR ELECTRONIC TECHNOLOGY, BRAIN, BODY, COGNITION RESEARCH AND BODY ART

Predrag Pale*
University of Zagreb, Faculty of Electrical Engineering and Computing, Zagreb, Croatia

Keywords: electronic technology, brain-body-cognition research, art, taxonomy, map

Extensive development of technologies, especially electronic, like sensors, wireless networks, signal processing, machine learning, and biomedical imaging significantly influences scientific research of brain, body and cognition. It also influences artistic explorations and expression offering them enhancement and multimodality of expression.

Objective: In order to leverage technologies, scientists and artists need insight into properties and limitations of technologies as well as projections for future development. Scientists and engineers developing new technologies, in turn, need overview of biomedical researchers' and artists' needs. Thus these three professions create a virtual developmental triangle in which each vertex is both dependent of the other two and provides valuable resources to them. The objective of the research presented in this paper is to facilitate improved understanding of all vertices in order to enable a more streamlined development of technologies and their leverage.

Method: A theoretical research exercise was undertaken to identify properties and limitations of technologies; the needs of artists that could be matched with technologies and the needs of biomedical researchers. The research was focused on present moment needs and abilities but significant effort has been undertaken to predict the future development. This is important since applications are always lagging the technology offerings for a few years.

Results: The results are a taxonomy of technology, a taxonomy of artists' needs and taxonomy of biomedical researchers' needs. A mapping among them is also provided.

Conclusion: By the mean of mapped taxonomies, engineers developing new technologies could get better insight into the users' needs which could influence the course of their decision on future development. Brain, body and cognition researchers could have much better overview and guide in choosing technologies and therefore foster their research. Just like researchers, artists too could benefit from mapped taxonomies by both leveraging technology and providing valuable input about their needs to engineers in charge of developing technologies.

In: Movement 2019: Brain, Body and Cognition
Editors: Joav Merrick and Gerry Leisman ISBN: 978-1-53616-714-6

Chapter 191

TRANSFERRING DRIVER TRAINING AND SAFETY RESEARCH IN DECISION MAKING UNDER STRESS, INTO CONTROLLING IMPULSIVITY AND INATTENTION IN MENTAL HEALTH REMEDIATION FOR GENERAL POPULATION

Abraham-Gerard Meyer*
Carnegie Mellon Driver Training and Safety Institute and MindfulChoice Laboratory, Pittsburgh, Pennsylvania, United States of America

Keywords: destination, attitude, focus, awareness, choice

Objective: Removing human a ways to think POSITIVELY; and form remote location, facilitating their own choice to decide on their individual journey to progress. Making every days few mini steps in the right direction for six to eight weeks – Until they satisfactory master the knowledge, and behavioral skills for doing it alone as their new way to proceed.

Methods: Prior any intervention, we are using online sophisticated automated screening to detect eventual co-morbid preconditions (ADHD, ODD, and CD, plus any of Procrastination's seven factors) – in order to customize for each patient a meta-cognitive positive psychology protocol for mind focused coaching and therapy - to remediate tunnel vision effect (TVE), a physiological dysfunctional responses triggering dire psychological and traumatic consequences associated with panic, obsession, anxiety, depression, PTSD. Constantly reinforcing; "Think good and it will be good".

Results: Positive noticeable outcomes show the success of a systemic binary approach, for populations of all ages - with a two prong education and training for both body (brain) and mind (intellect) health and wellness peak performance. This helps improving the metabolism to ensure a better flow of fresh oxygen in the brain, increase resistance to high stress, and consequently reducing negative thoughts and feelings, mood disturbances, impulsivity, inattention, aggressively, procrastination, and other negative emotional responses. General population's results are confirming our laboratory discovery that "Finally we end-up where we focus (physically, as well as mentally)".

Conclusion: We witnessed consistent and resilient positive changes that are visibly happening for all types of ages, to successfully stimulate general population. Using online face to face innovative visual and conceptual tools, as well as advanced technological methods, and processes for delivering to all clients meta-cognitive positive psychology education and training. It is unequivocally a quicker synergistic transformational avenue to follow for mental health professionals, and other therapists.

In: Movement 2019: Brain, Body and Cognition
Editors: Joav Merrick and Gerry Leisman
ISBN: 978-1-53616-714-6

Chapter 192

WRITING AN EFFECTIVE CASE REPORT FOR PUBLICATION

Seema Biswas**, *Chloe Pinto and Erez Shir On
British Medical Journal Case Reports (BMJ Case Reports), London, United Kingdom

Keywords: patient-based learning, global health

Objective: This workshop illustrates the importance of medical writing, shows new and reluctant authors, in particular, how to write effectively, and how to use global health case reports to advocate for better health and improved access to health care for their patients.

Methods: Published BMJ Case Reports will be used as examples to illustrate the importance of the patient at the centre of medical education, what may be learned from common conditions with unusual presentations, clinical dilemmas, the role and limitations of established clinical protocols, near misses, adverse events and initial cases of novel treatment regimens. Using BMJ Case Reports templates, participants will practice writing their own clinical case reports. In addition to describing the treatment of pathology, participants will be shown how to identify the determinants of health and disease in their patients and write global health case reports.

Results: By the end of the workshop, participants should be able to identify the learning value of their own clinical encounters and how to share this learning through the publication of a case report. Participants should be able to appreciate the role of global health case reports as tools to advocate for the resources they need to treat their patients and influence the conditions in which their patients live and work.

Conclusion: Case reports form an excellent first step in learning to prepare a manuscript. Learning from patients is essential to improving clinical care. The determinants of health are the causes of the causes of disease; influencing these determinants mitigates the health risks not only of individual patients but also of communities and populations. Analysis of global health issues in the preparation of global health case reports serves both to educate and advocate for improvement in the determinants of health.

In: Movement 2019: Brain, Body and Cognition
Editors: Joav Merrick and Gerry Leisman
ISBN: 978-1-53616-714-6

Chapter 193

SPASTICITY IS RELATED TO PRESSURE ULCER PREVALENCE IN ADVANCED DEMENTIA PATIENTS

Jaul Efraim*, Hagai Factor, Sharon Karni, Tehilla Schiffmiller and Oded Meiron
Department of Geriatric Skilled Nursing, Herzog Medical Center, Jerusalem, Israel

Keywords: spasticity, pressure ulcer, dementia

Objective: To identify and characterize the association relationship among prevalence of pressure ulcers, spasticity levels, and advanced dementia in disable elderly patients.

Design: The design was observational and cross sectional. Data was allocated from the medical files of elderly-immobile hospitalized patients, median age 73.5 years.

Setting: Patients were hospitalized in the skilled nursing departments at Herzog Medical Center, Jerusalem Israel.

Participants: 40 frail elderly patients bedbound, and suffering from advanced chronic diseases, advanced dementia and high-grade pressure ulcers.

Main outcome measure: Spasticity levels of all patients were rated according to the Modified Ashworth Scale. Pressure ulcer grades were scored according to the National Pressure Ulcers Advisory Panel (NPUAP) guidelines, and advanced dementia was evaluated using the Functional Assessment Staging (FAST).

Results: Logistic regression indicated that only advanced dementia and spasticity were significantly associated with development of pressure ulcers versus those without dementia or without spasticity. Patients with advanced dementia displayed significantly higher prevalence of severe spasticity. In a multivariate logistic regression analyses, spasticity was significantly associated with pressure ulcers, (Adjusted odds ratio = 11.033, 95% CI = 1.406-86.597; $p < 0.05$).

Conclusion: The current findings support significant association among the prevalence of pressure ulcers, spasticity, and advanced dementia, highlighting spasticity as the most significant risk factor associated with the development of pressure ulcers. Spasticity was significantly prevalent in advanced dementia patients versus non-demented patients supporting a reliable indication of rapid disease progression in dementia patients. The strong association of spasticity with the onset of pressure ulcers should encourage clinicians to systematically examine spasticity in this population in order to implement preventative measures, and delay the onset of pressure ulcers.

In: Movement 2019: Brain, Body and Cognition
Editors: Joav Merrick and Gerry Leisman
ISBN: 978-1-53616-714-6

Chapter 194

RESEARCHING THE FELDENKRAIS METHOD THROUGH PRINCIPLES OF JUDO

*Moti Nativ**
Bujinkan Dojo of Awareness (Shiki), Tel Aviv, Israel

Keywords: Feldenkrais method, awareness through movement, judo

Objective: Judo concepts and techniques had a significant impact on the development of the Feldenkrais method of somatic education. I researched Moshe Feldenkrais' activities as a martial artist in the years 1921-1950, digging into the depth of his work as a learner and a teacher of martial arts. This historical research aimed to investigate the influence of the martial arts on the development of the Feldenkrais method. As a martial artist, my study was not just theoretical. It involved experiencing the techniques myself and teaching them to others. A unique aspect of Feldenkrais' mastery of judo was his consistent attempt to analyze judo principles in terms of physics explanations and scientific reasoning. This intellectual development was also foundational for the application of judo principles in developing the Feldenkrais method. In this presentation I shall analyze my process of researching the relation between judo and the Feldenkrais method as a means of reflecting on researching the Feldenkrais method more generally.

Methods: Research was conducted through review of Feldenkrais publications, books, articles and recordings, and archival research (for instance – Bown judo collection in Beth University in England, Ben Gurion Archives, National Archive in Jerusalem, British national Archive), interviews with people that worked with Feldenkrais, and experiential testing of martial arts techniques.

Results: My historical research reveals that judo was the main environment for Feldenkrais to practically explore his ideas of awareness through movement. Moshe guided the judoka, coaching according to principles that we now know as the fundamental theories of learning according to the Feldenkrais method. Those can be summarized: 1) Start learning with slow movements. 2) Attain reversibility in action. 3) Attain good posture for freedom of movement and breathing. 4) Reduce effort to increase sensitivity (Weber-Fechner law). 5) Make mistakes.

Conclusion: Moshe's judo had a significant influence on the development of the Feldenkrais Method, and the fundamental principles of the Feldenkrais method had significant influence on Moshe's judo. Researching the Feldenkrais method requires both theoretical and experiential understanding of the martial arts roots of the Feldenkrais method, and presents a paradigm shift for methodologies researching the field of somatic education.

In: Movement 2019: Brain, Body and Cognition
Editors: Joav Merrick and Gerry Leisman
ISBN: 978-1-53616-714-6

Chapter 195

FAKE NEWS IN NEUROLOGY

Amos D Korczyn*
Tel-Aviv University Medical School, Ramat Aviv, Israel

Keywords: fake news, neurology

Fake news reached center stage in the last years through the messages of world leaders, but has in fact been around since the beginning of history, and constitute part of folklore. Their danger in politics is great and obvious, but it is also an important part of our everyday life in science and medicine, where its effects may be of great relevance.

Fake news, the deliberate spread of misleading information, occurs in the medical literature. One of its origins is the "publish or perish" culture whereby researchers are tempted to submit articles which contain unfounded claims. These may occasionally be discovered and withdrawn, and retracted papers probably only reflect a minority of such cases. Here is an important function of the goalkeepers, journal editors and the advisory boards and particularly the reviewers, to prevent publishing such articles. Publishing them may give them apparent credence.

Fake news is frequently intentionally spread in the non-scientific literature by interested parties, such as manufacturers aided by advertisers. These days social media provides fertile ground for the spread of unfounded medical claims about the successes of "miracle drugs". It is an important function of regulatory authorities to monitor such false claims and react to them, and of the medical community to be alert to exaggerated health claims and protect patients from exploitation and exposures to treatments which are frequently costly and potentially dangerous.

In: Movement 2019: Brain, Body and Cognition
Editors: Joav Merrick and Gerry Leisman
ISBN: 978-1-53616-714-6

Chapter 196

VALIDITY, RELIABILITY, INTERCLASS AND INTRACLASS CORRELATIONS AND CHARACTERISTICS OF NORMS OF THE ZUR BALANCE SCALE

Oz Zur[1,2], Matan Ravid[2], Bar Aviv[2], Roy Ravid[2] and Eli Carmeli[3,*]
[1]The Israeli Center for Dizziness and Balance Disorders, Raanana,
[2]Department of Physical Therapy, Ben-Gurion University in the Negev, Beer-Sheva
[3]Department of Physical Therapy, University of Haifa, Mt Carmel, Haifa, Israel

Keywords: balance, balance tests, function

Background: There are many balance tests, but ceiling effects, long time to administer and insufficient head movements are common disadvantages. In addition, no balance test examines the dynamic aspects and functional ability of the visual, vestibular and somatosensory systems, while standing.

Objective: This cross-sectional, double blind study was conducted to assess the validity, reliability, interclass and intraclass correlations of the clinical Zur Balance Scale (ZBS) and to determine the range of normal scores among different age groups.

Methods: A total of 140 employees of the company attended a frontal lecture about the purpose and procedure of the study, and among them, 88 signed consent forms to participate in this study, yet only 71 participants who completed all study examinations and met the inclusion and exclusion criteria were enrolled and included. In addition, 207 healthy friends and family participants of nine physiotherapists were recruited after providing signed informed consent. 251 healthy subjects (71 from a high-tech company and 180 friends of students), ages 29-70 years, were recruited and evaluated using the

ZBS. ZBS is a 4-5 minute test that evaluates 10 different balance conditions including head movements under different stance positions.

Results: ZBS had high inter-tester reliability (Cronbach's alpha 0.945, ICC 0.945), and high intra-tester reliability (Cronbach's alpha 0.938). ZBS scores decreased as age increased. It was found that the clinical ZBS is a sensitive test to detect balance disorders.

Conclusion: ZBS is simple and quick to administer, without a ceiling effect.

In: Movement 2019: Brain, Body and Cognition
Editors: Joav Merrick and Gerry Leisman
ISBN: 978-1-53616-714-6

Chapter 197

GIVING THEM WINGS: KEYS TO INTEGRATED DYNAMIC SHOULDER FUNCTION IN REHABILITATION

Joanne Elphinston*
JEMS Clinical Movement Education, Cardiff, United Kingdom

Keywords: shoulder, rehabilitation, brain-body integration

Objective: To introduce participants to an integrated approach which addresses the shoulder by understanding its motor patterns as modifiable behaviours, and in doing so, influencing its biomechanical and neuromuscular performance through simple perceptual and sensory techniques.

Description: Endlessly versatile, our shoulders present a paradox: our extravagant mobility is made possible through a compromise in structural stability, yet when performing well, our shoulders are capable of significant functional stability even in extreme positions. We can perform extraordinary feats of strength and impressive force transference through our upper limbs in large dynamic movements, yet we are able to caress the strings of a violin or remove a splinter from a child's finger with exceptional delicacy and dexterity.

We can only achieve this with finely coordinated global integration, a shifting landscape of neuromuscular partnerships and a dance between our biomechanics and myofascial system. Our emotions, beliefs and self-efficacy also have significant impact on the way in which we use our upper limbs, influencing the outcome of the movement before it has even begun, especially after injury or trauma.

This experiential workshop opens a window on these relationships, exploring the power of our verbal and tactile cuing to alter global biomechanics and neuromuscular patterning; providing simple, immediately applicable techniques which modify movement by changing perception; sensory motor techniques which establish global connections; and whole body integration experiences to balance shoulder loading across the kinetic chain.

The workshop challenges the cognitive traditions of "mind over matter" which reflect a divided self, instead reflecting more contemporary science by drawing on the individual's own sensory experience, clarifying and modifying intention in their movement, and harvesting their positive emotional associations to alter predictive motor processing.

Ultimately, if we get beyond muscles and ask instead, "what does the brain think this movement is about?" we discover the master key to facilitating change in upper limb function.

In: Movement 2019: Brain, Body and Cognition
Editors: Joav Merrick and Gerry Leisman
ISBN: 978-1-53616-714-6

Chapter 198

A FUNNY THING HAPPENED ON THE WAY TO THE CONFERENCE: PERCEPTION OF HUMOR IN PARKINSON'S DISEASE

Jenni Posen, Yael Manor, Avner Thaler, Clara Mayanz, Dana Yekutiel, Nir Giladi and Tanya Gurevich*
Movement Disorders Unit, Department of Neurology, Tel Aviv Medical Center and Sackler School of Medicine, Tel Aviv University, Tel Aviv, Israel

Keywords: Parkinson's disease, humor, MoCa

Background: Situations and content perceived as humoristic may help patients cope with chronic disease ramifications and symptoms such as pain, discomfort and depression. Frontal and temporal lobes in the brain and the limbic system are involved in humor comprehension and are also areas damaged in Parkinson's disease (PD). As a first step towards utilizing humor as a coping tool in a PD population, we are conducting an ongoing study on its perception by PD patients in comparison with a healthy control.

Aim: To characterize perception of different types of humoristic content by people with PD.

Method: A study of 7 PD patients (5 men, 2 women, age 63.7 ± 9.3 education years15.7 ± 2.36, illness duration 7.8 ± 4.8,) and 7 healthy control (5 women, 2 men, age 57.6 ± 6.8, average education years 19 ± 2.44), not demented according to Montreal Cognitive Assessment (MoCa) > 26. Participants underwent Unified Parkinson's Disease Rating (UPDRS), cognitive screening (Frontal Assessment Battery (FAB) + MoCa) and completed questionnaires: Geriatric Depression Scale (GDS 15), Spielberger Trait –State Anxiety Inventory, Parkinson's disease quality of life (PDQ 8), Sense of Humor

Questionnaire (SHQ- 6). Patients were also asked to rate as humorous 3 nonverbal cartoons with punch lines, 3 audio 3 visual presentations, portraying different humor styles (blatant, subtle, macabre).

Results: No significant differences were found in FAB and MoCa scores. Significant differences were demonstrated in GDS15 education and age. Significant differences in appreciation of humoristic content were demonstrated in 2/3 video presentations ($p = 0.04$, $p = 0.002$). No significant differences were noted in audio and cartoon presentations.

Conclusion: PD patients experienced more difficulties in perceiving video content as humoristic than healthy controls probably due to the attention load required in watching videos. It appears a less effective medium for utilizing humor in PD. Although age and education might also impact.

In: Movement 2019: Brain, Body and Cognition
Editors: Joav Merrick and Gerry Leisman
ISBN: 978-1-53616-714-6

Chapter 199

READOUT OF KINEMATIC INFORMATION IN AUTISM EXPERTS

Novella Pretti[1,2,*], Alessandro Crippa[3], Andrea Cavallo[1,2], Francesca Battaglia[4], Caterina Ansuini[1], Massimo Molteni[3] and Cristina Becchio[1]

[1]C'MoN Cognition, Motion and Neuroscience Unit,
Fondazione Istituto Italiano di Tecnologia, Genova, Italy
[2]Department of Psychology, University of Turin, Torino
[3]Child Psychopathology Unit, Scientific Institute, IRCCS Eugenio Medea, Bosisio Parini
[4]Child Neuropsychiatric Unit, G Gaslini Institute, Genova, Italy

Keywords: autism, kinematics, readout, clinical expertise

Objective: Over the past 20 years, retrospective coding of home videos by experienced examiners has proved useful for assessing qualitative aspects of sensorimotor behavior early in the development of autism spectrum disorder (ASD). However, the specific sensorimotor features detected by experienced examiners are unknown. Are examiners with extensive experience in the assessment of ASD children able to detect features invisible to non-experts? The present study was designed to provide quantitative answers to this question. Specifically, we evaluated the extent to which expert examiners are able to pick-up significant information from subtle differences in movement kinematics.

Methods: Participants were clinical examiners with extensive experience in the assessment of ASD children (experts) and clinical examiners with no training in the

* Corresponding Author's Email: novella.pretti@iit.it.

assessment of ASD children (non-experts). All participants completed two tasks: an autism discrimination task and an intention discrimination task. Each task consisted of a series of trials displaying reach-to-grasp actions towards a bottle performed by TD and ASD children either with the intent to pour or place. In the autism discrimination task, we required participants to indicate whether the observed action was performed by a TD or an ASD child. In the intention discrimination task, we required participants to predict whether the observed action was performed with the intent to pour or place. For each task, we presented 120 videos selected to contain both autism-specifying information and intention-specifying information.

Results: Data collection is ongoing. Piloting with control participants with no clinical expertise indicates task-related differences in the ability to pick-up subtle changes in movement kinematics. Specifically, control participants are able to use kinematic information to discriminate intention, but not autism. We expect that expert examiners will outperform non-experts (and control participants) in the autism discrimination task, but not in the intention task.

Conclusion: To the best of our knowledge, this is the first study to examine the influence of clinical expertise on video analysis in the field of autism research. Our results demonstrate the potential significance of kinematic readout for clinical expertise and suggest that examiners with extensive experience in assessing ASD develop a task-specific ability to pick-up ASD-diagnostic kinematic features.

SECTION TWO: ACKNOWLEDGMENTS

In: Movement 2019: Brain, Body and Cognition
Editors: Joav Merrick and Gerry Leisman
ISBN: 978-1-53616-714-6

Chapter 200

About the Authors

Joav Merrick, MD, MMedSci, DMSc, born and educated in Denmark is professor of pediatrics, Division of Pediatrics, Hadassah Hebrew University Medical Center, Mt Scopus Campus, Jerusalem, Israel and Kentucky Children's Hospital, University of Kentucky, Lexington, Kentucky United States and professor of public health at the Center for Healthy Development, School of Public Health, Georgia State University, Atlanta, United States, the former medical director of the Health Services, Division for Intellectual and Developmental Disabilities, Ministry of Social Affairs and Social Services, Jerusalem, the founder and director of the National Institute of Child Health and Human Development in Israel. Numerous publications in the field of pediatrics, child health and human development, rehabilitation, intellectual disability, disability, health, welfare, abuse, advocacy, quality of life and prevention. Received the Peter Sabroe Child Award for outstanding work on behalf of Danish Children in 1985 and the International LEGO-Prize ("The Children's Nobel Prize") for an extraordinary contribution towards improvement in child welfare and well-being in 1987. In 2017 appointed a Kentucky Colonel by the Commonwealth of Kentucky, the highest honor the governor can bestow to a person. Email: jmerrick@zahav.net.il

Gerry Leisman, MD, PhD, is an Israeli neuroscientist educated in the United Kingdom and the United States in Medicine, Neuroscience, and Biomedical Engineering at Manchester University. He additionally received a PhD in Neuroscience and Biomedical Engineering from Union University in 1979. He is currently Full Professor and Research Fellow at the University of Haifa in Israel. He is also concurrently Director of the National Institute for Brain and Rehabilitation Sciences, a global research institute with laboratories in Gilbert, Arizona, United States, Nazareth and Zevulun, Israel and in Havana, Cuba. He concurrently is Professor of Restorative Neurology at the University of Medical Sciences in Havana, Cuba. He is Editor-in-Chief of the journal *Brain Behavior Cognition*.

His scientific career has largely been involved since the early 1970s in the promotion of consciousness as a scientifically tractable problem, and has been particularly influential in arguing that a fundamental understanding of consciousness can be approached using the modern tools of neurobiology and understood by mechanisms of theoretical physics, having, together with Dr. Paul Koch, developed biomedical applications of continuum theory. He has also been influential in examining mechanisms of self-organizing systems in the brain and nervous system applied to cognitive functions exemplified by his work in movement-cognitive interaction, memory, kinesiology, optimization, consciousness, death, autism, and developmental disabilities. He has likewise applied optimization strategies to movement and gait, cognition, and coma recovery. It is in this context that he was one of the first to identify functional disconnectivities in the brain and nervous system. He was elected Fellow of the Association for Psychological Science in 1990, a Senior Member of the Engineering in Medicine and Biology Society of in IEEE in 1986, and a Life Fellow of the International Association of Functional Neurology and Rehabilitation in 2010, having received its Lifetime Achievement Award in 2011. He has served as a consultant to the US National Sciences Foundation.

He is the co-author of the text with Dr Robert Melillo, *Neurodevelopmental disorders: An evolutionary perspective*, with Dr Joav Merrick on *Considering consciousness clinically*, and on *Neuroplasticity in learning and rehabilitation*. He has published hundreds of papers in the neurosciences, cognitive science, biomedical engineering, and in the systems sciences literature. Email: g.leisman@edu.haifa.ac.il.

In: Movement 2019: Brain, Body and Cognition
Editors: Joav Merrick and Gerry Leisman
ISBN: 978-1-53616-714-6

Chapter 201

About the National Institute of Child Health and Human Development in Israel

The National Institute of Child Health and Human Development (NICHD) in Israel was established in 1998 as an independent virtual institute under the auspices of the Medical Director, Ministry of Social Affairs and Social Services in order to function as the research arm for the Office of the Medical Director. In 1998 the National Council for Child Health and Pediatrics, Ministry of Health and in 1999 the Director General and Deputy Director General of the Ministry of Health endorsed the establishment of the NICHD. In 2018 the NICHD became affiliated with the Center for Children with Chronic Diseases, Division of Pediatrics, Hadassah Hebrew University Medical Center, Mt Scopus Campus, Jerusalem, Israel.

Mission

The mission of a National Institute for Child Health and Human Development in Israel is to provide an academic focal point for the scholarly interdisciplinary study of child life, health, public health, welfare, disability, rehabilitation, intellectual disability and related aspects of human development. This mission includes research, teaching, clinical work, information and public service activities in the field of child health and human development.

Service and Academic Activities

Over the years many activities became focused in the south of Israel due to collaboration with various professionals at the Faculty of Health Sciences (FOHS) at the Ben Gurion

University of the Negev (BGU). Since 2000 an affiliation with the Zusman Child Development Center at the Pediatric Division of Soroka University Medical Center has resulted in collaboration around the establishment of the Down Syndrome Clinic at that center. In 2002 a full course on "Disability" was established at the Recanati School for Allied Professions in the Community, FOHS, BGU and in 2005 collaboration was started with the Primary Care Unit of the faculty and disability became part of the master of public health course on "Children and society". In the academic year 2005-2006 a one semester course on "Aging with disability" was started as part of the Master of Science program in gerontology in our collaboration with the Center for Multidisciplinary Research in Aging. In 2010 collaborations with the Division of Pediatrics, Hadassah Hebrew University Medical Center, Jerusalem, Israel around the National Down Syndrome Center and teaching students and residents about intellectual and developmental disabilities as part of their training at this campus.

RESEARCH ACTIVITIES

The affiliated staff have over the years published work from projects and research activities in this national and international collaboration. In the year 2000 the International Journal of Adolescent Medicine and Health and in 2005 the International Journal on Disability and Human Development of De Gruyter Publishing House (Berlin and New York) were affiliated with the National Institute of Child Health and Human Development. From 2008 also the International Journal of Child Health and Human Development (Nova Science, New York), the International Journal of Child and Adolescent Health (Nova Science) and the Journal of Pain Management (Nova Science) affiliated and from 2009 the International Public Health Journal (Nova Science) and Journal of Alternative Medicine Research (Nova Science). All peer-reviewed international journals.

NATIONAL COLLABORATIONS

Nationally the NICHD works in collaboration with the Faculty of Health Sciences, Ben Gurion University of the Negev; Department of Physical Therapy, Sackler School of Medicine, Tel Aviv University; Autism Center, Assaf HaRofeh Medical Center; National Rett and PKU Centers at Chaim Sheba Medical Center, Tel HaShomer; Department of Physiotherapy, Haifa University; Department of Education, Bar Ilan University, Ramat Gan, Faculty of Social Sciences and Health Sciences; College of Judea and Samaria in Ariel and in 2011 affiliation with Center for Pediatric Chronic Diseases and National

Center for Down Syndrome, Department of Pediatrics, Hadassah Hebrew University Medical Center, Mount Scopus Campus, Jerusalem.

International collaborations

Internationally with the Department of Disability and Human Development, College of Applied Health Sciences, University of Illinois at Chicago; Strong Center for Developmental Disabilities, Golisano Children's Hospital at Strong, University of Rochester School of Medicine and Dentistry, New York; Centre on Intellectual Disabilities, University of Albany, New York; Centre for Chronic Disease Prevention and Control, Health Canada, Ottawa; Chandler Medical Center and Children's Hospital, Kentucky Children's Hospital, Section of Adolescent Medicine, University of Kentucky, Lexington; Chronic Disease Prevention and Control Research Center, Baylor College of Medicine, Houston, Texas; Division of Neuroscience, Department of Psychiatry, Columbia University, New York; Institute for the Study of Disadvantage and Disability, Atlanta; Center for Autism and Related Disorders, Department Psychiatry, Children's Hospital Boston, Boston; Department of Pediatric and Adolescent Medicine, Western Michigan University Homer Stryker MD School of Medicine, Kalamazoo, Michigan, United States; Department of Paediatrics, Child Health and Adolescent Medicine, Children's Hospital at Westmead, Westmead, Australia; International Centre for the Study of Occupational and Mental Health, Düsseldorf, Germany; Centre for Advanced Studies in Nursing, Department of General Practice and Primary Care, University of Aberdeen, Aberdeen, United Kingdom; Quality of Life Research Center, Copenhagen, Denmark; Nordic School of Public Health, Gottenburg, Sweden, Scandinavian Institute of Quality of Working Life, Oslo, Norway; The Department of Applied Social Sciences (APSS) of The Hong Kong Polytechnic University Hong Kong.

Targets

Our focus is on research, international collaborations, clinical work, teaching and policy in health, disability and human development and to establish the NICHD as a permanent institute in Israel in order to conduct model research and provide policy aspects in the field of interest.

CONTACT

Professor Joav Merrick, MD, MMedSci, DMSc
Director, National Institute of Child Health and Human Development, Jerusalem, Israel
Email: jmerrick@zahav.net.il.

In: Movement 2019: Brain, Body and Cognition
Editors: Joav Merrick and Gerry Leisman
ISBN: 978-1-53616-714-6

Chapter 202

About the The Institute for Brain and Rehabilittion Sciences

The Institute for Brain and Rehabilitation Sciences is a research institution founded in 2010 in the State of Israel and operates in Nazareth and Zevulun. Its primary mission is to serve as a one-stop-shop for all matters that relate to rehabilitation sciences and includes research programs in biomechanics, ergonomics, imaging sciences, translational neurosciences, trauma recovery, developmental disabilities, nutritional sciences and immunology, ergonomics and all matters that allow for an individual to return to work and support him or herself after life-altering injury.

The Institute, based in Israel, has formal relationships with the National Institute for Brain and Rehabilitation Sciences in the USA, Haifa University, the M.I.N.D. Institute at M.I.T., The University of the Medical Sciences and the University of Havana both in Cuba, and Bielfeld University in Germany and has published in excess of 100 papers in collaboration with these institutions. Additionally, the Institute produces a journal called *Brain Behavior and Cognition*, now in its 9th volume. A 14 volume Handbook of Rehabilitation Sciences is also in process.

The Institute is active in the incubation of patentable ideas and as such has a portfolio of patients and technologies with applications ranging from the early detection of PTSD, to novel ways of reducing health risks in the Fourth Age, artificial implantable dialyzers and many other potential life affirming products in rehabilitation applications.

The Institute's philosophy is life affirming and as such makes no distinction between education and rehabilitation. The focus is less on the concept of disability and disorder and more on the nature of how to optimize human performance. As such, the Institute is involved with developing the field of Neuroeducation and has published widely on the subject in both the scientific, clinical and public policy journals. Institute representative have lectured at Oxford University on the role of the Neurosciences in fomenting changes in classroom learning. The Institute has presently ongoing collaborative research programs in Neuroeducation with DePaul University in Chicago USA.

Underneath all of the Institute's work is the active study of the human connectome. It is how it is that brain regions communicate with each other that serves as the basis of life and death, effective living and learning, the facilitation of the generation of brain substances that promote neuroplasticity that are achieved by motor activity and all that can support the promotion of cognitive and physical function across the lifespan. Applications of studies in this area have allowed for the elucidation of the nature of Persistent Vegetative States, Neurobehavioral disorders of childhood including autism spectrum disorders and have even found linkages between infectious diseases and autism with the Institute's researchers having provided testimony for government and insurance company funding decisions for conditions such as P.A.N.D.A.S. (Pediatric autoimmune neuropsychiatric disorders associated with streptococcal infections).

Contact

Professor Gerry Leisman
University of Haifa,
Faculty of Social Welfare and Health Sciences
Abba Khoushy Ave. 199, Eshkol Tower, Rm. 910
Mt Carmel, Haifa, Israel 3498838
Email: g.leisman@edu.haifa.ac.il
Website: www.nibrs.org

In: Movement 2019: Brain, Body and Cognition
Editors: Joav Merrick and Gerry Leisman
ISBN: 978-1-53616-714-6

Chapter 203

About the Book Series "Functional Neurology"

This is a book series with publications from a multidisciplinary group of researchers, practitioners and clinicians for an international professional forum interested in the broad spectrum of functional neurology, health and human development. Book series editors: Gerry Leisman and Joav Merrick.

- Leisman G, Merrick J, eds. Considering consciousness clinically. New York: Nova Science, 2016.
- Leisman G, Merrick J, eds. Neuroplasticity in learning and rehabilitation. New York: Nova Science, 2016.
- Vojdani A. Neuroimmunity and the brain-gut connection. New York: Nova Science, 2016.
- Merrick J, Leisman G, eds. Movement 2018: Brain, body and cognition. New York: Nova Science, 2018.

Contact

Professor Joav Merrick, MD, MMedSci, DMSc.
Director, National Institute of Child Health and Human Development,
Jerusalem, Israel.
Email: jmerrick@zahav.net.il.

SECTION THREE: INDEX

INDEX

A

D

E

F

L

M

N

O

P

T

U

V

W

Y